BIOLOGY
OF MAMMALS

Biology
of Mammals

🐾 🐾 🐾 🐾 🐾

RICHARD G. VAN GELDER

CHAIRMAN AND ASSOCIATE CURATOR,
DEPARTMENT OF MAMMALOGY,
THE AMERICAN MUSEUM
OF NATURAL HISTORY

CHARLES SCRIBNER'S SONS · NEW YORK

COPYRIGHT © 1969 RICHARD G. VAN GELDER

This book published simultaneously in the
United States of America and in Canada—
Copyright under the Berne Convention

All rights reserved. No part of this book
may be reproduced in any form without the
permission of Charles Scribner's Sons.

A-6.69 [C]

Printed in the United States of America

Library of Congress Catalog Card Number 68-27782

For Rosalind, Russell, and Gordon,
who helped to complete
my evolutionary purpose

🐾 🐾 🐾 PREFACE

The lives of mammals are so complex and so individualistic that no single volume could adequately cover all of the topics. More than 100,000 books and papers have been written about mammals, and the volume of literature grows at a rate of 2 to 3 per cent a year. Despite all these works, relatively little is known of the lives of most species. Even those that are considered to have been well investigated have more facets of their lives unknown than known. That there should be great variety in the lives of mammals is understandable from the viewpoint of evolutionary doctrine, for it is from variations that natural selection weeds out those unsuited to the environment.

The idea of this book was to outline some of the ways in which mammals vary to adapt to the stresses of environment. I have drawn mainly on my own knowledge of North American and South American mammals and the literature to ask "why" and "how." I have tried to locate underlying principles essential to the lives of mammals and to generalize from them. By generalizing I have undoubtedly compounded errors and certainly ignored most exceptions. However, both the generalizations and exceptions are excellent stimuli to further thought, and it is hoped that they will stimulate more profound examination of these subjects by others.

PREFACE

This volume is deliberately brief, for it was planned only as an annotated outline. Coverage has been limited to those topics that I considered essential or to those that were of especial interest to me.

I have been fortunate in having available the excellent library facilities of The American Museum of Natural History and the Department of Mammalogy. Even more valuable has been the benefit of many provocative discussions with my colleagues at the American Museum: Sydney Anderson, Karl F. Koopman, Guy G. Musser, and Hobart M. Van Deusen. Helene Jamieson Waddell of The Rockefeller University was of great help in preparing the initial outline for this book. Cecile M. Cusson typed the manuscript with her customary precision and accuracy and offered much helpful advice. Russell Neil Van Gelder provided me with numerous examples of elemental mammalian behavior, and Gordon Mark Van Gelder later confirmed these. My wife, Rosalind, read the entire manuscript and freely provided suggestions and encouragement throughout its progress. To all of these persons I am deeply and heartily grateful.

RICHARD G. VAN GELDER

CONTENTS

PREFACE *vii*

1 · Introduction *1*
2 · Birth, Growth, Development *14*
3 · Dispersal *27*
4 · Home Range, Territory, Shelter *50*
5 · Air, Water, Food *62*
6 · Defense and Protection *85*
7 · Social Structure and Populations *104*
8 · Mating, Reproduction, Gestation *123*
9 · The Value of Mammals *141*

APPENDIX *161*
REFERENCE NOTES *171*
GLOSSARY *175*
SELECTED BIBLIOGRAPHY *183*
INDEX *191*

1 ❦ Introduction

AMONG the million or more kinds of animals that inhabit the earth today, the mammals are a small group, comprising less than 0.5 per cent of the total. They are relative newcomers in the evolutionary history of the earth, and their peak in evolution, as represented by number and diversity of kinds, has probably passed. Compared with the majority of other kinds of animals, however, mammals are relatively large and some species are quite numerous, so that, in total, they exert upon the environment an influence that is disproportionately great in relation to their four thousand species.

The mammals are grouped together with the sea squirts, lancelets, fishes, amphibians, reptiles, and birds into an evolutionary unit known as the Phylum Chordata. This group has as its unifying characters distinctive anatomical features that function in support of the body, nervous transmission, and breathing. Although some of these characters cannot be seen in the adult chordates, evidence of common evolutionary heritage is found in the structure of their developing embryos. The vertebral column, or backbone, of mammals allies them structurally with a subdivision of the chordates, the Subphylum Vertebrata, which also includes fishes, amphibians, reptiles, and birds. It is within the Subphylum Vertebrata that the mammals evolved from the reptiles.

The fishes are a group of vertebrates that are tied to water for their way of life. No fish can live the entire cycle of its life, from birth to production of young, away from an aquatic environment. The amphibians—frogs, toads, and salamanders—have become more independent of the water, but generally return to it to deposit their eggs or require a moist medium for their development.

The reptiles, by evolutionary perfection of egg shells and a scaly skin to protect against desiccation, became the first group of vertebrates to become strongly independent of an aquatic environment. It is from the reptiles that the mammals arose.

The lines of evolution that led to the mammals had their origin early in the history of the reptiles (Late Pennsylvanian), although the mammals themselves became distinguished relatively late (Late Triassic–Early Jurassic). Among the changes that appear in the fossils are some that have become characters of modern mammals. There has been a reduction in the number of bones of the skull and jaws, culminating in mammals in the single bone that forms each side of the lower jaw. Differentiation of types of teeth for cutting, stabbing, grinding, crushing, and shearing, characteristic of mammals, also appears in the mammal-like reptiles. There were also many changes in the skeleton of the body and limbs, some in the direction of increasing running ability.

But the main characters by which we distinguish modern mammals are not evident in fossils. In concise terms, mammals could be described as lactogenic, hirsute homoiotherms—milk-producing, hairy, warm-blooded animals—and it is seldom that fossils show the structures that would give evidence for these facets of physiology and morphology.

Of the differences between reptiles and mammals, homoiothermism is undoubtedly the most important. However, be-

INTRODUCTION

cause it is a characteristic shared with the birds and therefore not unique to mammals, its significance is often underestimated. Yet, almost every aspect of life that makes a mammal different from a reptile is dependent, in one way or another, upon this fact that mammals are able to maintain a relatively high and relatively constant body temperature by their metabolic activity. (Many reptiles are also capable of maintaining a relatively high and constant body temperature, but accomplish this, mainly, by absorbing heat from the environment or sheltering from excessive heat.)

When active, most mammals maintain a body temperature between 95–100.4°F. The body covering of hair serves as insulation against both colder and hotter air temperatures, but when the temperature exceeds the insulating ability of the fur, other mechanisms come into play. Panting, salivation, and sweating, plus dilation of the blood vessels in the skin, help the animal to lose heat and thereby to reduce body temperature. In cold, the metabolic rate is increased, the blood vessels in the skin constrict, and muscular activity (even in the form of shivering) may serve to maintain the internal body temperature at a level high above that of the external environment. The Arctic fox, with its fine insulating fur, can endure temperatures of $-22°F$. without having to increase its metabolism over the basal (resting) rate.

Such characteristics of mammals as non-nucleate red blood cells (increased surface for absorption of oxygen), a four-chambered heart (no mixture of oxygenated and unoxygenated blood, so that a richer supply of oxygen for metabolism is in the blood reaching the tissues), and hair (for insulation) are obviously closely associated with thermoregulation and homoiothermism.

The milk production of mammals for the nourishment of their young is also largely dependent upon homoiothermism.

The complex glandular relationships of mammals depend extensively upon a uniform temperature for their action. Lower temperatures slow the speed of chemical reactions, and temperatures only a few degrees higher than the normal body temperature may cause breakdown of the endocrine system.

Nourishment of the young with milk, and the consequent dependence of the young upon the mother, has led to another specialty of the mammals, parental care. While such nurture is typical also of birds, it is not a characteristic of the reptiles, whose young are born or hatched well developed and receive virtually no attention from their mother.

The comparatively long period of dependence of the young upon the mother in mammals has been coordinated in many of them with a greater development of the cerebral portion of the brain. Thus, the suckling period has also become a time of learning for many of the mammals, and a most important part of their lives. In this cerebral development and the degree of ability to learn, mammals differ from all other animals.

CLASSIFICATION

The mammals are divided into groups based on similarity of structure and common heritage. All mammals are placed in a single class, the Class Mammalia, which is one of at least five classes of living vertebrates. To the specialist there are a multitude of subdivisions within the class, but in this book the names of most of these are not given, and the most uniform level used is that known as order. Of the living mammals, eighteen orders are commonly recognized. The names of the orders, with some comments about each, follow.

ORDER MONOTREMATA, the monotremes. About six species,

placed in two families, are in this order. These are the platypus and the echidnas, which are found only in Australia, Tasmania, and New Guinea. Monotremes are the only living mammals that lay eggs. They possibly represent an independent crossing of the reptile-mammal line and thus may be less closely related to all the other mammals than are the other orders.

ORDER MARSUPIALIA, the marsupials. There are about 242 species, placed in about nine families, in this order. South America and the Australian region are their main areas of occurrence. Although they are often called "pouched mammals," not all of them have a pouch (and the echidna, a monotreme, does have a sort of pouch). The young are born in a relatively undeveloped stage and continue their growth attached to the mother's nipples which, in the pouched species, are within the pouch. The opossum is the only marsupial that ranges north of Mexico. Some of the Australian marsupials are kangaroos, wallabies, phalangers, bandicoots, wombats, and the koala.

Placentals

Mammals in all of the following orders are called placentals because of their extensive development of connections (the placenta) between the embryo and the mother so that the developing animal receives nourishment from the adult while it is still in the uterus. This is not generally true in the marsupials. At the present time it is thought that the marsupials and placentals are derived from a common ancestor, but that their lines of evolution diverged early in the history of these two groups.

ORDER INSECTIVORA, the insectivores. Some four hundred

species (probably less) in eight families make up this order, which is considered by most mammalogists to represent the group from which the other placental orders arose. The insectivores are found in most parts of the world except Australia and much of South America. They are relatively small animals and include the shrews, hedgehogs, and moles. As their name implies, most feed on insects and other invertebrates.

ORDER DERMOPTERA, the dermopterans. There are only two species, in a single family, in this order. They are commonly called "flying lemurs," colugos, or kaguans. They are not lemurs, however, and they do not fly. They are adapted for gliding, having extensive membranes of skin from the neck to the toes, from front to hind toes, and enclosing the tail. They are found only in the Philippine Islands and southeastern Asia, Sumatra, Java, Borneo, and adjacent islands.

ORDER CHIROPTERA, the bats. This is the second largest order of mammals, with some 875 species in 16 families. Bats are the only mammals capable of true flight, which is accomplished by wings formed of skin between the elongate fingers and the body. Bats are found on all continents except Antarctica, north or south to the limit of trees.

ORDER PRIMATES, the primates. There are about 166 species in 10 families in this order, which includes man, the apes, monkeys, marmosets, lorises, tarsiers, and lemurs. They are mainly tropical-forest animals of Africa, Asia, and America. Man is a notable exception to this distribution.

ORDER EDENTATA, the edentates. About 31 species in three families comprise this strictly American group. Although the scientific name of this order implies that they are toothless, only one family, the anteaters, actually lacks teeth. The other families in this order are those of the sloths and of the armadillos.

ORDER PHOLIDOTA, the pangolins. There are eight species

INTRODUCTION

of pangolins, all in a single family. They are found only in tropical Africa and southeastern Asia. Pangolins are toothless and covered with scales.

ORDER LAGOMORPHA, the lagomorphs. About 63 species in two families make up this order, which was originally world-wide in distribution, except for Australia where man has now introduced them. The pikas of North America and Asia, the rabbits, and the hares, make up this order.

ORDER RODENTIA, the rodents. This is the largest of all the orders, having 34 families and about 1,687 species. They are world-wide in distribution, and the order includes squirrels, rats, mice, beavers, muskrats, chinchillas, hamsters, porcupines, gerbils, gophers, and cavies.

ORDER CETACEA, the cetaceans. The members of this order, the whales, porpoises, and dolphins, are wholly aquatic. There are about 84 species in 10 families, and they are found in all oceans and in some rivers and lakes. Some taxonomists divide this order into two: the toothed whales and the baleen whales. The order includes the largest of all mammals, the blue whale.

ORDER CARNIVORA, the carnivores. Ten families and about 284 species make up this order, which is world-wide in distribution, except for the Australian region; there dogs were introduced by man. For the most part this group is composed of flesh eaters. Three families (seals, sea lions, and walrus) have sometimes been placed in a separate order, Pinnipedia. Some members of this group are the bears, dogs, wolves, foxes, weasels, skunks, badgers, otters, mongooses, pandas, raccoons, kinkajou, ferrets, genets, civets, wolverine, lion, tiger, leopards, cheetah, lynx, and other cats, and hyenas.

ORDER TUBULIDENTATA, the aardvark. There is only one family and one species in this order. The aardvark is an African termite eater with a tubular snout.

ORDER PROBOSCIDEA, the elephants. Two species in a

single family comprise the living elephants. They are found in Africa and southern Asia. The African elephant is the largest living land mammal.

ORDER HYRACOIDEA, the hyraxes. There are 11 species of hyraxes, all in one family. They are found in Africa and the Near East and superficially resemble pikas.

ORDER SIRENIA, the sirenians. Only four species in two families of this order survive today. A fifth species, Steller's sea cow, was exterminated in the eighteenth century. The manatees and dugongs are wholly aquatic animals that inhabit tropical and subtropical coasts and some rivers.

ORDER PERISSODACTYLA, the perissodactyls. There are 16 species in three families in this order. These hoofed mammals have the axis of the foot along the midline of the middle toe, and the members of this order are sometimes referred to as the "odd-toed" hoofed animals, because they generally have one or three toes on each foot. The horses, zebras, asses, tapirs, and rhinoceroses are in this order, which is native to both the Old and New Worlds.

ORDER ARTIODACTYLA, the artiodactyls. These are the "even-toed" hoofed or cloven-hoofed mammals; the axis of the foot lies between the third and fourth toes. There are about 171 species in nine families. Pigs, peccaries, hippopotamuses, camels, llamas, deer, giraffes, antelopes, cattle, sheep, and goats are all members of this order, which is native throughout the world except for the Australian region.

The orders and families are further subdivided into smaller groups of common heritage. The genus (plural, genera) and the species are subdivisions of the family. The species is the basic unit of classification and is composed of animals that, in nature, may interbreed. From other species, any given species is reproductively isolated. For example, the coyote

(*Canis latrans*), the timber wolf (*Canis lupus*), the lion (*Leo leo*), and the cheetah (*Acinonyx jubatus*) are all in the Order Carnivora. The coyote and wolf are members of a single family, Canidae, while the lion and cheetah are in a different family, Felidae. Both the coyote and the wolf belong to the same genus, *Canis,* while the two cats are in different genera, *Leo* and *Acinonyx*. Although they are in the same genus, timber wolves and coyotes do not, in nature, interbreed, and because of anatomical differences as well as the reproductive isolation, are placed in different species. The scientific name is made up of the generic and specific names, so that *Canis lupus* is the timber wolf and *Canis latrans* is the coyote. Although these two wild dogs are somewhat similar in general appearance and anatomy, which is partially why they are placed in the same genus, the lion and cheetah are quite different anatomically and in appearance, and they do do not interbreed. Not only are they placed in different species, but also in different genera, the lion in the genus *Leo* (to which the tiger, jaguar, and leopard also belong) and the cheetah in the genus *Acinonyx*, of which it is the only living species.

Of course, not all members of a species are identical, and in different parts of the range of a species, populations of animals have evolved slightly differently. Those populations that can be distinguished from others of the same species sometimes receive a third scientific name, the subspecies. Thus, the coyote from the southern part of the peninsula of Baja California, Mexico, is known as *Canis latrans peninsulae,* and it can be distinguished from the one from the northern part of the peninsula, which is known as *Canis latrans clepticus*. The two can interbreed where they come in contact, so they are not different species, but the two populations are distinguishable, so they are subspecies.

GEOGRAPHY OF MAMMALS

The distribution of mammals on earth is roughly from the North Pole to the coastal shelf of Antarctica, both on land and in water. There is no major type of habitat that is not inhabited by some type of mammal, and because of their thermoregulation, cold is less a general limiting factor in their distribution than it is for many cold-blooded animals.

The whales, porpoises, and dolphins are wholly oceanic mammals, spending their entire lives at sea. The sirenians are also highly adapted for life in water, but they are generally inhabitants of coastal waters, estuaries, or rivers, because they are herbivorous. Seals, sea lions, and walruses are less restricted to water than are the sirenians and cetaceans, for they are capable of movement on land and generally give birth to their young on shore. These littoral species, however, may venture far to sea at certain times of the year. Sea otters, also, are essentially littoral species, feeding on invertebrates from the floor of the sea.

Many of the orders of mammals have aquatic representatives that shelter or feed in fresh-water streams or lakes. Species adapted for fresh-water activity are found in the monotremes (platypus), marsupials (yapok), insectivores (*Potomogale*), carnivores (otters), rodents (nutria, beaver), lagomorphs (marsh and swamp rabbits), cetaceans (Amazon dolphin), sirenians (Amazon manatee), and ungulates (hippopotamus).

Of the mammals that have invaded the air, only the bats are capable of true flight, that is, they can sustain themselves in the air. However, several kinds of rodents have achieved locomotion through the air in the form of gliding (flying squirrels), and the same is true for some marsupials (gliding possums). Both species of the Order Dermoptera are highly specialized for gliding by the development of skin membranes.

The majority of mammals are terrestrial. Some types are highly specialized for a burrowing, or fossorial, existence (moles, pocket gophers), while others are less specialized but dig burrows for protection or for storage of food while spending most of their active time on the surface of the ground (woodchuck, echidna). Most of the species that spend their time on the surface of the ground are quadrupedal, although a few (kangaroos, kangaroo rats, jerboas, man) may move bipedally.

Trees are inhabited by a variety of mammals in several orders. Sharp claws (tree squirrels), hooked claws (sloths), grasping hands (some primates), and prehensile tails (some South American monkeys, some marsupials, the kinkajou, and pangolins) are some adaptations of specialized arboreal mammals. As already mentioned, gliding has been achieved by a few arboreal species.

ZOOGEOGRAPHIC REGIONS

As the major parts of the world were explored and as general knowledge of the distribution of animals was acquired, scientists became aware that certain parts of the world seemed to have a fauna peculiar to the particular region. This led, in the nineteenth century, to a descriptive division of the world into zoogeographic regions. There are various subdivisions within each grouping, but in broad, general terms, the generally recognized zoogeographic regions are the following:

Australian Region. This area includes Australia, Tasmania, New Guinea, and some adjacent islands. Monotremes are found only in this region and it is also the home of many kinds of marsupials.

Oriental Region. Southeastern Asia, south of the Himalayas, and including Sumatra, Borneo, Java, Celebes, and the Philippine Islands make up this zoogeographic area. Gibbons, the orangutan, and tarsiers are among the mammals limited to this region.

Ethiopian Region. Africa south of 20 degrees north latitude makes up this zoogeographic area, which also includes Madagascar. The gorilla, chimpanzees, the giraffe, hippopotamuses, many kinds of antelopes, and (in Madagascar) lemurs are typical.

Neotropical Region. This area includes all of South America and Central America north to central Mexico. Sloths, anteaters, prehensile-tailed monkeys, guinea pigs, and the llama are characteristic mammals.

Nearctic Region. North America from central Mexico northward and Greenland comprise this region. Prong-horned antelope, mountain goat, and muskrats are examples of Nearctic mammals.

Palearctic Region. This region is composed of all of Eurasia north of the Himalayas and 20 degrees north latitude in Africa. Hedgehogs, fallow deer, and roe deer are examples of Palearctic mammals.

Holarctic Realm. The Nearctic and Palearctic are the least differentiated of the major zoogeographic regions, and they are sometimes united into a single realm under the name Holarctic. Some Holarctic species are mountain sheep, bison, moose, red fox, and lynx.

In summary, the mammals are backboned, warm-blooded, haired animals that suckle their young. There are about 4,000 species living today, and these are classified in 18 orders.

INTRODUCTION

Mammals are world-wide in distribution, inhabiting both land and sea, from the North Pole to the coasts of Antarctica.

In the following chapters aspects of the lives of mammals will be discussed, and examples of the way various mammals cope with problems of survival will be given.

2 ❧ Birth, Growth, Development

BIRTH

The life of any mammal is part of an evolutionary continuum. Each species goes through a cycle of life that is completed when the offspring of the next generation reach the same point on the cycle. From the evolutionary point of view it seems logical to start this outline of the biology of mammals with birth and to follow the natural cycle through growth, development, dispersal, adult life and social behavior, and to complete it with reproduction of the next generation before terminating with some reflections on the relationships of mammals to their environment and to man.

Birth is both a beginning and an end. For the newborn mammal it is the beginning of independent life, and for the parent it is an end to the gestation period. The period of prenatal care for the young ends, and postnatal care begins. Birth is also an end in itself—for with the birth of its offspring the parent has fulfilled an evolutionary mission—to reproduce itself.

Birth takes place in several ways in mammals. The young

platypus or echidna hatches from an egg. The baby platypus uses a hard protuberance known as an egg tooth to cut open the leathery shell and emerges naked, blind, and practically helpless into the nest. Having fulfilled its purpose, the egg tooth disappears after birth.[1]* Echidnas are probably born in a similar fashion, except that the egg is incubated in the mother's pouch and the young hatches there.

Aside from these two egg layers, all other mammals are viviparous. They give birth to living young that vary in amount of development according to species. Some are naked, blind, and nearly helpless at birth, while others are well developed and able to move about with the parent soon after being born.

The marsupials give birth to little more than a partly developed fetus, which continues its development while attached to a nipple, usually in a pouch. Compared with the adult, and relative to other mammals, the newborn marsupial is tiny. The American opossum, about as large as a domestic cat, has young that are the size of a bee at birth; even in the great gray kangaroo, which may be six feet tall, the young are only an inch long at birth. The forelimbs of the newborn marsupial are well developed, and with them it pulls itself through the mother's fur to find its way to a teat. Generally, the mother makes no effort to help the baby find her nipple, but on a few occasions kangaroos have been seen to assist the young with their lips and paws. The infant nonpouched marsupials, such as murine opossums, merely hold firmly onto a nipple and are carried about in this fashion; the mother elevates her hind quarters when walking to keep the young from dragging.

All the rest of the mammals give birth to somewhat better-developed young that have received nourishment through the placenta while within the mother. Although birth has been observed in relatively few wild animals, it seems to take place most commonly with the mother lying down. Animals with

---

* Reference Notes are grouped in the back of the book.

certain adaptations for locomotion or habitat may, however, give birth in other positions. Some bats assume an inverted quadrupedal position on the roof of a cave and bear the young into a "basket" formed by the membranes that join the legs and tail; other bats give birth while hanging in their normal head-down position. The agouti, in contrast to other rodents, and the chimpanzee have been seen to give birth while in a squatting position, and a gibbon was observed to give birth while hanging from a branch. Whales, porpoises, the dugong, and manatees bear their young in the water, as do the hippopotamuses. The newborn manatee is carried above the water's surface on the back of its mother for 45 minutes, and during the following two hours she gradually submerges it into its normal, aquatic environment.

In mammals that ordinarily bear a single offspring at each birth, the young are generally born head first. This is true of most hoofed animals and primates. However, breech presentation—tail first—is more common in the whales, where parturition may last as long as twenty minutes; a young animal that has partially emerged head first might readily drown. Head presentation is less common in the rodents and carnivores, which generally have several relatively small young at a time.

WELL-DEVELOPED AND POORLY DEVELOPED YOUNG

Newborn placental mammals are precocial (well developed) or altricial (poorly developed), although there are variations in each category. Precocial mammals are born with the eyes open and with abundant hair; they are capable of moving about shortly after birth and often can eat some solid food soon after birth. Precocial young generally are characteristic of animals without nests or dens that live in open country where speed is often the main means of escape. Most of the hoofed animals—deer, cattle, goats, and sheep—as well as elephants, whales, porpoises, hares, and some rodents (porcupines, agoutis, and guinea pigs) have more or less precocial young.

Altricial offspring are usually naked and blind at birth, and they require a relatively long period of parental attention before they have grown sufficiently to fend for themselves. Precocial offspring are not immediately independent of parental care, but they are born at a stage that is reached by altricial young only after some weeks of additional development. Most primates, rodents, rabbits, carnivores, bats, and insectivores have altricial young.

MILK

After hatching or birth the newborn mammal's nutrition, in the form of milk, still comes from the mother. There is a relationship between degree of development of the young at birth, the nutritive value of the milk, and the consequent rate of growth of the young animal. Milk is composed of water, fats, proteins, sugars, and various mineral salts in proportions that vary with the species. Those mammals that produce milk with the highest protein content have young that grow the fastest; calcium and phosphorus also are present in relatively high proportions in the milk of some mammals with especially fast-growing young.

Milk is produced by the mammary glands, which are specialized skin glands. Early in the embryonic development of a mammal, milk lines, two thickened lines of skin, extend from axilla to groin, and it is along these lines that the mammae ultimately form. The teats' position on the milk line and their number vary according to the species as well as within a given species—in marsupials, for instance, the number may be anywhere between two and twenty-seven—although the number is generally more or less constant for a group. Primates, bats, elephants, manatees, sloths, and some South American rodents have only pectoral mammae (in the chest area); many hoofed animals and the whales and porpoises have only inguinal teats (in the region of the groin).[2]

Animals with localized teats usually suckle their young standing. The elephant, with pectoral teats, and deer, goats, cattle, and sheep, whose teats are inguinal, exemplify such mammals. Those with several pairs of mammae along the milk line—dogs, cats, many rodents, and pigs—generally lie down to permit the young to nurse.

The actual secretion of milk by the mother is stimulated at birth by a lactogenic hormone (known as galactogen or prolactin) which comes from the pituitary gland located at the bottom of the brain; adrenal hormones also are essential to lactation. During gestation the production of lactogenic hormone from the pituitary is inhibited by the presence of another hormone, estrogen, which disappears at birth. In some mammals the stimulus of the sucking offspring serves, through the nervous system, to stimulate the liberation of lactogen.

The platypus and echidnas are teatless, and their young do not begin to nurse for several days after they have hatched. Milk production in these mammals seems to be stimulated by the young's nuzzling the belly of the mother with the tip of the bill, and the milk is excreted from specialized sweat glands onto the skin of the abdomen. These glands are associated with bunches of hair from which the milk is lapped by the babies. In the echidnas these glandular patches are in the pouch. Marsupial young are attached firmly to the nipples, and the nipple swells in the mouth so that the young can scarcely be detached without tearing tissues. The windpipe of the babies is prolonged and abuts the end of the nasal passage so that they can breathe while suckling. In the pouched marsupials, the nipples are in the pouch.

In the placentals the young nurse until they are sated, generally a matter of some minutes. Whales and porpoises are faced with a special problem when nursing their young—to provide them with milk, yet permit them to breathe. This is accomplished by giving the infant a large quantity of milk in a few seconds: the touching of the nipple by the young porpoise or whale causes a contraction of the muscles surrounding the mammary gland, thus squirting milk into the baby's mouth.

WEANING

Few data are available concerning the length of the suckling period in wild animals. Weaning generally occurs somewhat later than the time when the young actually can eat solid food. A nursing period of a week and a half to three weeks is general for the smaller rodents. Hares, although they may eat green food less than two weeks after birth, nurse for a month. Porcupines can eat solid food in the first month, but they are suckled for four or five weeks. Elk calves can eat green food from the age of two weeks but may nurse for six or more months, while some other deer, which nibble vegetation in a few days, are weaned in six weeks. Moose may nurse for nearly a year and reindeer almost as long. Elephants suckle their young for at least two and sometimes five years, walruses for two years. Bison may nurse for a year, and whales for six or seven months; a captive bottlenose dolphin, which ate bits of fish and squid at six months, was nursed for a year and a half. Seals may nurse less than two months, and the harp seal—

whose young weigh from twelve to fifteen pounds at birth and grow within two weeks to sixty to eighty pounds—deserts the infant after two or three weeks. The young seal remains on an ice floe for a week or two after the mother has left, and then takes to the water, first eating small crustaceans and later fish.

GROWTH

The rate at which mammals grow varies with the species and depends in part on the degree of development of the young at birth, the quality of the mother's milk, the length of the nursing period, the number of litter mates, the life span of the species, and many other factors. Because the weight of an animal is generally proportional to its volume, weight is often used to measure growth, because it is more readily obtained than linear dimensions. If the weight of the animal is plotted against time on a graph, a sigmoid (S-shaped) curve generally results. This shape curve illustrates that actual size increases slowly before birth, rapidly after the baby is born, and tapers off as adult dimensions are reached. To obtain the relative growth rate, the amount of weight gained during constant time periods is recorded as a percentage of the birth weight. The curve's peak indicates that the most rapid weight increments occur when the animal is very young.

The conditions under which a species of mammal lives make certain demands upon the growth rate of the young. An Arctic hibernator, such as the Arctic ground squirrel, must attain a sufficient amount of weight, including fat, within fewer than five months after birth in order to survive a seven-month hibernation period. By the time the young ground squirrels are a month and a half old they are four-fifths the size of adults and have attained almost three-fifths of their adult weight; at four months of age, just before going into hibernation, they have attained adult size and weight and have enough fat to last them through hibernation.

Short-lived species may have a very rapid growth rate. Long-tailed shrews, which rarely live more than a year and a half, weigh 1/200 ounce at birth but have attained two-thirds of the weight of an adult (1/12 to 1/2 ounce, depending on the species) by the time they are weaned at three weeks of age. Contrarily, however, some of the bats of the family Vespertilionidae are weaned and independent in four weeks, but they may live to twenty or more years of age. The slow-growing African elephant does not reach adult size until it is twenty, although it may live more than fifty years.

Sexual maturity—that is, the physiological ability to breed—is generally attained before adult size is reached. In many mammals, especially rodents, one or both sexes may breed as subadults; in other kinds of mammals the subadults, although potentially capable, do not breed until adulthood is reached. There are various reasons for this; among them are the social behavior of the species, environmental conditions prevailing, and population size.

SIZE OF MAMMALS

An animal's rate of growth tends to decrease with age. Although growth is always taking place in the form of replacement of cells and the formation of blood and antibodies, the ultimate size of an animal is limited by the size of its bones when it is adult, and the cessation of bone growth is largely under genetic and glandular control. An animal is supported by its bones; a doubling of the linear dimensions—length, height, and width—of an animal increases the volume by the product of the three measurements. Thus the weight increases cubically. At the same time the area of the supporting bones increases by the product of two dimensions. An animal that doubles its linear dimensions has squared the supportive area of its bones and has cubed its weight. But, for bones to support the cubed weight that results from a doubling in size, an animal would have to cube the cross-section area of its long bones. This would probably make the animal so cumbersome that it would scarcely be able to move on land. Gravity thus tends to limit the size of land mammals. Modern giraffes may reach 20 feet in height and 12 feet at the shoulder, but weigh no more than 2 tons; 10 tons is near the limit for the largest living terrestrial mammal, the African elephant. Some extinct mammals were larger; the rhinoceroslike *Paraceratherium,* which lived in Mongolia, is the largest land mammal that is known ever to have existed; it stood about 18 feet at the shoulder.[3]

Marine mammals, freed from much of the effect of gravity through the support of water, reach the largest sizes ever known for a living creature. Blue whales have exceeded 100 feet in length and 150 tons in weight, or more than twenty times the weight of the average circus elephant. So needed is this aqueous support that large whales that are inadvertently stranded on beaches usually die from suffocation caused by the pressure of their own great bulk pressing on their lungs. What limit there is to the size of oceanic mammals is unknown.

In addition to being limited in maximum size, land mammals also have a minimum that, strangely enough, is controlled by food. As mentioned earlier, mammals are homoiothermic—warm-blooded. Although there are variations of this physiological state, it is temperature regulation that seems to govern the lower size limit of a mammal. The smallest living mammals are some of the shrews; the Etruscan shrew and the pigmy shrew each weigh 2.3 grams—197 individual shrews to the pound.

The smaller the volume of the animal, the greater is its proportional surface area. For example, a cube 1 inch on a side has, in proportion to its volume, twice as much surface area as one that is 2 inches on a side. (The 1-inch cube has a surface area of 6 square inches and a volume of 1 cubic inch. The 2-inch cube has a surface area of 24 square inches and a volume of 8 cubic inches—this is a proportion of 1 to 6 for the 1-inch cube as opposed to only 1 to 3 for the 2-inch one.) Heat loss in a warm-blooded animal takes place through the exposed surfaces, and in order to obtain sufficient food to maintain their high body temperature, shrews must be active almost constantly. Bats that are nearly as small as shrews avoid the problem of heat loss when they are inactive by not controlling their body temperature and permitting it to follow, more or less, the fluctuations of the air temperature.

During the period of most rapid growth, the young mammal develops in many morphological, physiological, and behavioral ways. Depending, of course, on the species involved, hair is developed, the eyes open, muscular coordination is developed and improved, thermoregulation is developed, milk teeth are lost, hunting techniques are learned, juvenile pelage is molted, and weaning takes place.

Fox squirrels, for example, born naked and with their eyes closed, develop a fine coat of hair in ten days but do not open their eyes for another month. It is still another two weeks before muscular coordination is sufficiently well developed for the young to move from the nest. Excepting those mammals that are precocial at birth, the newborn lack the ability to regulate their temperature and in their early life are essentially heterothermic, or cold blooded. In nature, of course, warmth is provided by the mother's heat, the insulating qualities of the nest, or through other environmental factors, but some young altricial mammals show remarkable resistance to cold. Norway rats, for example, younger than ten days of age can survive freezing temperatures and even a complete lack of oxygen for a comparatively long time. Even the human offspring is physiologically a cold-blooded animal for the first few months of its life.[4]

It is often during the preweaning period that the deciduous dentition (milk dentition) is lost. Mammals are diphyodont, that is, they have two sets of teeth in the course of their lives, excepting the whales, which are monophyodont, having only a single set of teeth, and baleen whales have teeth only while they are embryos.

The milk teeth are the precursors of the permanent incisors, canines, and premolars. (In the marsupials only the last premolars are preceded by deciduous teeth.) There are no deciduous precursors of the permanent molars. While bats and a few other mammals lose their milk teeth before birth (*in utero*), most others lose them at various preweaning ages and some long after they have been weaned.

Among those species in which the color or texture of the pelage of the young is markedly different from that of the adult, the change usually takes place at about the time the animal is able to care for itself, even though it may remain with the mother for a much longer time.

The food for growth and development is provided for baby mammals by the parents. Most young mammals are cared for by the mother alone, but the family life of very few species is well known. Among some kinds of mice the male may share in the repair of the nest or burrow and even help to transport the babies. In the carnivores, such as wolves, foxes, and weasels, it is not uncommon for the male to bring food back for the family. Males of gregarious species may be the main defenders of the herd, and giraffes, for example, maintain a sort of "baby-sitting" arrangement in which a few adults may look after all of the young of a certain age.

DEVELOPMENT OF BEHAVIOR

As physical and physiological growth proceed, development of behavior also occurs. Behavior patterns are either innate (instinctive) or learned. Comparatively little is known of either innate or learned behavior development in wild mammals. These processes are related to the physical and physiological chronology of development.

Innate behavior probably is modified by subsequent learning through trial and error or by demonstration by the parent. However, many innate processes are brought into play at stages in life when the animal has no one to teach it. Nest and den construction and sexual behavior are primarily innate behavior patterns. On the other hand, for some animals defensive reactions appear to be learned. Some young rats, born in captivity, showed no fear of a large snake, while their parents, caught when wild, attacked the snake violently, presumably having learned to defend themselves against snakes in their native Africa. Many other young animals born in captivity show a similar ignorance when compared with wild-caught adults. Otters and some seals must force their young into the water and help them while they are learning to swim.

A large number of behavioral processes are learned as young animals play. Fawns of various species of deer have races (learning to run fast to escape enemies, or learning to run with the herd). Young predators will chase a rolling pebble or snap at a drifting feather, perhaps learning to catch prey. Mock fights—biting and wrestling—are common and may be methods of learning defense.

As proficiency is gained in the various physical, physiological, and psychological activities of an animal, its dependency on its parents decreases, and it finally is ready to care for itself. Most mammals are capable of an independent existence somewhat before they actually undertake it, either on their own or when induced to by a parent.

3 ❧ Dispersal

REASONS FOR DISPERSAL

Lack of space is one of the main causes of dispersal. As young animals approach adult size, mere physical crowding in the den may necessitate that some move out. Competition, however, is not limited to need for space alone, and most other biological factors are also involved. With the approach of another breeding season, social competition may also require dispersal, for males may recognize their offspring as such only up to a certain point, and after sexual maturity is reached, the young are additional competitors for mates.

A given area can support only a certain number of animals of one species, and the presence of additional adult individuals, with their concomitant demands on the food supply and den sites, is another reason for dispersal. With animals that may have a second litter while the previous one is still in a den of limited size, there is danger of cannibalism or physical injury to the youngest animals.

Among animals that have homes or limited individual ranges, the young are usually gone from the site of birth before the appearance of a second litter. For example, young coyotes, born in the spring, begin to wander away in the autumn, perhaps encouraged in this by rebuffs from the parents when youngsters try to return to the home den. There may be an inherent urge to wander in the young of some species so that, as they reach adult size, they merely drift away from home. This seems to be the case with the North American raccoon.

In other species, one of the parents may actually drive the young away. In an American beaver family, the young remain with the parents almost two years, but just before the appearance of the third litter, the two-year-old animals are driven out of the home by the mother. Even the male, for the beavers seem to mate for life, may have to move out of the lodge temporarily before the appearance of the litter, leaving the mother and a set of yearlings to await the new arrivals.

The young of many other mammals are deserted by the mother about the time they are reaching maturity or are capable of caring for themselves: The mother of the black-tailed prairie dog moves out, leaving the burrow to the young, and goes to live in a deserted burrow or digs a new one. A white-footed mouse female, expecting a second litter, may desert her almost-grown young and move to a new nest. Eastern gray squirrel females expecting a second summer litter will also move away from the first batch of youngsters, although the second litter will probably stay with the mother through the winter. In cases when there is no second litter, the female squirrel may be driven from the home nest by her young.

As mentioned earlier, the young of some seals, such as the harp seals, are nursed only a few weeks and then are deserted by the mother, who returns to the sea to mate and to feed. The infants may remain on the ice for several weeks, often molting during this period, before venturing into the water to commence feeding. Alaskan fur-seal females desert their young to migrate southward.

Mammals that do not have fixed homes and that move about over a large area in search of food do not have the need to disperse when the young have grown. In mammals that form herds, especially the hoofed animals of grasslands and savannas, the young simply join the group; this is probably the case with many antelopes and wild cattle. Some segregation may take place within the herds, however, as juvenile groups of a single sex or age class form. Young Alaskan fur-seal males, which may not be able to begin assembling a harem until they are six years of age or older, gather together in bachelor herds at the rear of the rookery, being more or less forced there by the adult bull harem masters that will allow no trespass of the young males into their territories.

A similar situation exists among the vicuñas of the Andes of southern South America. Adult males, accompanied by several females and their young, have territories that they defend against other such bands. Yearling females move out of the band at the time their mothers are having young and join the bands of other adult males. The yearling males, however, move from their "home" bands at this same time and form male troops of leaderless young vicuñas, mostly consisting of yearlings and two-year-olds.[1]

HAZARDS OF DISPERSAL

The period of dispersal is one of the most hazardous in the life of a mammal. At this time it is no longer subject to parental care and protection, but still is a young, relatively inexperienced animal, just beginning to face new situations in which it must learn to survive. Those young, dispersing mammals that do not quickly learn soon die. Most of the striped skunks trapped during the autumn and winter are the young of the year, as are most of those killed on the highways during this period. The surviving skunks are either inherently more cautious or have learned to avoid traps and cars.

The dispersing juvenile may also venture into the territory of another member of its species. This will usually result in the wanderer's being chased off by the adult defender, and the young animal will continue to be exposed to the hazards of dispersal for a longer time. Physical fights in the defense of territory at this time are extremely rare.

Young animals unable to find suitable living sites may be forced into regions in which they or their offspring cannot survive, or in which they are unable to reproduce. Invasion of an unsuitable area is of no importance to the evolution and survival of the species if the population cannot be maintained.

Often the young disperse shortly before the winter, and therefore these less experienced young animals have their chances of survival reduced during this period of cold weather and short food supply. Even among animals that hibernate, to live through the winter the young must put on a sufficient amount of fat during the short period from the time they have left the parental domain until the time for hibernation. They must also find or construct for themselves a suitable hibernation site.

Scarcity of food supply may cause young animals to disperse from the parental territory even earlier than usual. This behavior has been noted in California ground squirrels, with a concomitant reduction in rate of survival.[2] While few of these adult squirrels, which live in an area where hibernation is a marginal requirement, are active above ground throughout the year, young animals are more apt to be out through the winter, or at least the first one. This would also tend to reduce the possible survival of the young should a cold snap or further reduction in food supply occur.

In general competition with other members of the species, young mammals tend to lose out to mature adults. Even among the young, there is some competition, and when dusky-footed woodrats are dispersing, the smallest individuals usually go farthest from the home, because the larger litter mates take the sites closest to the parental den. It has been noted that a young rat of this species has a better chance to survive if its mother disappears after it has been weaned and it is able to remain in the maternal home.[3]

IMPORTANCE OF DISPERSAL

Dispersal is of great importance in the lives of mammals. Genetically it is a means of preventing too much inbreeding and permits genes to spread throughout a species. It prevents the overpopulation of an area and an excessive drain on the food supply with its subsequent effect on vegetation and other food sources. It is a means of natural selection by which the better-adapted young animals have a greater chance to survive than do their less well-adapted siblings, and, of course, the constant exploration of new areas by the dispersing young is one way in which a species may populate a new region if it is capable of surviving there. Also, it is a means of population control, by providing animals to be eaten by predators on the one hand, while assuring continuation of the species because the adults of breeding age usually remain in tried and tested sites where survival may be better assured than in some unknown region.

LONGEVITY

If a young mammal leaves the parental home and establishes itself in a new territory, its chances of survival are enhanced. It still must face a multitude of mortality factors, and few wild animals ever live out their potential life span. Old animals may lack the speed to escape a predator or to catch their prey. Defensive ability may be reduced as reaction time slows. Their teeth may be worn down to the point where they can no longer chew their food, sight or hearing may fail, or their joints may become so stiff that they cannot reach their food. Disease resistance may be lowered. These animals die quickly. In captivity, however, the life span of animals may approach the potential maximum, because human care may permit them to survive despite infirmities. Most of the data available on the longevity of mammals come from records of captives.

Actually, adult animals have the best chance for survival. Some small mammals with naturally short lives have hardly any theoretical life expectancy at all. Adult deer mice, field mice, short-tailed shrews, and Norway rats, have a 95 per cent or higher probability of dying within a year. Adult rabbits and hares of several species have a 20 to 50 per cent expectancy of living a year, while in the larger mammals such as mule deer, roe deer, and caribou, the adults have a 58 to 75 per cent annual expectancy of living. For baleen whales the annual life expectancy for an adult is in the neighborhood of 66 to 78 per cent.

Life tables have been constructed for a few species of mammals, and these point up the times of high mortality rate. It is estimated that a population of 1,000 newly born cottontail rabbits has a theoretical average life expectancy of 6.5 months. However, in the first four months of their lives, 744 of these 1,000 newborn will die. The mortality rate, then, is 74.4 per cent. From the fourth to the fifth month, 28 of the 256 survivors will die. The mortality rate for this month is 11.4 per cent, much lower than the 74.4 for the newborn, and the potential life span for the survivors is 6.6 months. By the end of the first year, only 80 of the original 1,000 cottontails would have survived, and 18 per cent would be expected to die within a month. After 18 months, only 38 rabbits would be alive, with an expected mortality during the eighteenth to nineteenth months of life of 13 per cent. At 2 years, only 16 of the original 1,000 rabbits would be alive, and 8 per cent will die within a month. At 2½ years there would be 5 rabbits, with an expected mortality of 14 per cent. After 3 years, theoretically, only 0.4 of a rabbit would remain, and at 3¼ years, none of the original 1,000 would be alive.[4]

The longest-lived mammal is man, who may live more than a hundred years. In regions where medical knowledge and medication are not available, the life expectancy of a human being is much lower than where such assistance is available. In Egypt, for example, the 1958 infant mortality was as high as 145 per thousand births. But even in a place so advanced in medicine as the United States, the probability of death is greater for infants than for adults. For the entire U.S. population recently, there were 9.3 deaths per thousand persons, but among infants (mainly under one month of age) the rate ranged from a low of 17.4 per thousand in Nebraska to a high of 37.4 in Mississippi—two to four times as high respectively as the overall rates.[5]

The elephants, reputed to live hundreds of years, are not known to exceed 70 years of age, and probably only live about half this length of time in the wild. The domestic horse is close behind with a record life span of 62 years; an echidna has reached more than 50 years of age; a zoo hippopotamus, 49 years; and a rhinoceros, 47. Bears, lions, tapirs, domestic cattle, and, rarely, domestic dogs, may live into their thirties, while many kinds of deer, camels, domestic pigs, zebras, whales and dolphins, porcupines, gibbons, orangutans, baboons, rhesus monkeys, lemurs, and some bats are known to have lived into their twenties. Most of the smaller rodents and insectivores live only a few years at most. The life expectancy of any meadow mouse is less than a month, although individuals may survive a year and a half or more in the wild.

MOVEMENT

In order to disperse, as well as to find food or shelter, to escape from enemies, and to find mates, a mammal must be able to move. Although there are animals, even members of the Phylum Chordata, that are sessile (attached to one place) for most of their adult lives, there are no mammals that are not capable of locomotion. The manner in which mammals move, however, varies with the species and the speed at which the individual is moving.

Terrestrial Locomotion

For ground-dwelling mammals, walking is the slowest form of locomotion; there are different types of this motion among the four-legged ones. Most of the hoofed animals, members of the cat and dog family, and many rodents walk by moving one foot at a time. This means that the body is supported by the other three feet during walking, and the sequence of movement is always of a particular type. If the first limb moved is the left forefoot, the second moved will be the right hind foot, the next the right forefoot, and then the left hind foot, so that the sequential movement of the limbs is always diagonally across the body. When standing, quadrupedal mammals display two main types of posture. Horses are representative of a group that has the center of gravity nearer the forelegs. A horse can kick out with one hind leg, or a blacksmith may lift its hind foot without trouble. But for the animal to lift its foreleg, the weight must first be shifted onto the hind legs. Rabbits, squirrels, and bears have their center of balance closer toward the hind legs, and thus either forelimb can readily be lifted. Kangaroos have a center of gravity that is actually behind the hind limbs, and the animal, when erect, must use its tail as a brace to avoid falling over backward.

In a slow walk, then, a quadruped is always supported by three limbs and can stop in its tracks without toppling over. As it begins to move a little faster, the next limb in the sequence of movement is lifted before the previous one reaches the ground, and there is a short period during which only two feet—diagonally opposite ones—are on the ground, with a consequent loss of stability. In this manner, the gait has become a trot, in which the sequence of footsteps is the same as in a walk, but faster, and the lack of stability is overcome by the rapid placement of counterbalancing feet. For the greater speed achieved, the animal has sacrificed stability. In the gallop, the fastest of quadrupedal gaits, a horse moves its feet in the following sequence, left front foot, right front foot, left hind foot, right hind foot. Usually only one foot is touching the ground at any time, and there are even periods during which none of the feet is touching the ground. This gait has again sacrificed stability for speed. The gallop of a weasel is slightly different, both the forelimbs being moved together followed by the hind limbs, and hares and rabbits show a combination of the gallop of a weasel plus that of a horse; they move the hind limbs together and the front limbs separately.

Not all quadrupedal mammals employ the typical diagonal footwork, although the majority do. Elephants, giraffes, camels, hyenas, and some young dogs move both legs on the same side at the same time in a gait called pacing, or ambling. This imparts a rolling movement to the gait that has been known to make riders of camels and elephants motion-sick.

Amphibious mammals such as seals and sea lions are somewhat handicapped for terrestrial motion, but not as much as one would at first think. The sea lions and fur seals can walk and run on land because they are able to turn their hind limbs forward to support some of the body weight and have large front flippers that also support the body on land. Walruses also have foreflippers that are about a quarter the length of the body and move on land in the same fashion as the sea lions. Both sea lions and walruses are able to make long journeys overland; in stormy weather walruses have been reported to travel 15 or 20 miles over snow.

True, or hair, seals differ from sea lions and walruses in that they are incapable of turning the hind flippers forward and thus cannot use them in the conventional manner for terrestrial locomotion. Weddell's seals have such short front flippers that their main means of locomotion on ice floes is like that of an inchworm—an arching of the back and creeping. Seals with flippers that can reach the ground use these in locomotion, coupled with an arching of the back and drawing of the hindquarters forward, and then by straightening the back, which moves the front part of the body forward. Locomotion using mainly the forelimbs can produce a fair amount of speed, and the crab-eater seal has been recorded as moving over a half mile of hard-packed snow as fast as a sprinting man, up to 15 miles an hour. The seal accomplishes this by making powerful backward strokes with the front flippers and strong blows against the snow with its hind flippers.[6]

Not all land-dwelling mammals move on all fours, of course, but those that are bipedal are relatively few. Man is the only tail-less mammal that consistently and regularly uses a bipedal form of locomotion. Apes are capable of walking erect without touching the ground with their hands, but seldom do. Gibbons, which are essentially arboreal, can walk erect on the ground without using their front limbs for support, but they hold their arms either over the head or out to the sides for balance. Gerenuks, long-necked African antelopes, often rear on their hind legs to reach foliage out of their reach in the quadrupedal position. Other hoofed animals are known to do this as well, not only when feeding but also during the mating season when, for example, horses and wapiti may rear onto the hind legs to bite or to strike an opponent with the forefeet. Bears, too, can walk on their hind legs alone, but, again, this is not their usual form of locomotion.

A different form of bipedalism has been developed by a number of other animals in various parts of the world. It is generally associated with a long tail that is used for support or balance, or both. Leaping animals, such as kangaroos, jerboas, and kangaroo rats, propel themselves by means of the powerful hind legs. During the longest leaps of a kangaroo, which may cover more than 25 feet, the tail does not touch the ground, but is used entirely for balance. At a very slow pace, a kangaroo may use five points of contact with the ground, supporting the body with the small front legs and the tail while the hind legs are moved forward. Kangaroo rats and jerboas rarely use the forelegs in locomotion, and the latter easily cover a foot or two in their normal unhurried hops, 6 to 10 feet when chasing one another, and can leap up to 12 or 15 feet in a bound when severely pressed—a remarkable leap for an animal that measures only about 6 inches from tip of nose to base of tail and that has a tail almost 10 inches long and a weight of a third of a pound.

A few nonleapers with stout tails are able to use bipedal locomotion. The giant armadillo of the Amazon basin and the pangolins are examples of other animals that may occasionally use a bipedal, tail-supported gait. Spotted skunks, and, more rarely, striped skunks, may sometimes rise onto their forefeet and walk a few steps. However, this action is to warn enemies and is not a means of locomotion.

Arboreal Locomotion

Mammals that spend all or most of their lives in trees have developed adaptations that enhance this way of life for them. Arboreal locomotion is, in general, a modified form of terrestrial movement with the equivalent of the walk, trot, and gallop. A grasping hand or foot is one of the arboreal adaptations, most evident in the monkeys, and in some mice and marsupials. Another aid to arboreal life is the tail, which is used to balance the animal when it is walking on narrow limbs or making leaps. A prehensile tail, which is capable of grasping a tree limb for support, has been developed in a number of mammalian orders including many of the opossums and phalangers (Marsupialia), the kinkajou (Carnivora), South American arboreal anteaters (Edentata) and porcupines (Rodentia), some pangolins (Pholidota), and some New World monkeys (Primates).

Sharp, curved claws are an obvious aid to a climbing animal, and among the arboreal species, their development reaches its extreme in the sloths, which spend most of their lives hanging upside down from branches by their long, hooked claws. They move quite poorly on the ground, not even being able to stand erect, although they can swim well.

Another form of arboreal locomotion is found in some of the primates. Brachiation, as this is called, is essentially a form of bipedal locomotion that involves the forelimbs. As perfected by the gibbons, it involves grasping with one hand (with the body dangling), and, as the body swings forward, the momentum that carries it past the vertical position of the arm is utilized for forward motion by grasping with the other hand, as the body pivots through 180 degrees. Gibbons can move 10 feet in a swing in this way, and when leaping from one branch to another, they glide through the intervening space. Other primates, such as the South American spider monkeys, also use brachiation.

Aquatic Locomotion

Many mammals that live on land are capable of swimming. Bats, sloths, and moles can swim, and most rodents, ungulates, and carnivores also are good swimmers. In fact, relatively few mammals cannot swim, among these being chimpanzees, gorillas, and gibbons. Some mammals too heavy to stay afloat, such as armadillos, still can cross water. The nine-banded armadillo can walk across the bottom of a small stream or it can swim. To increase its buoyancy it is able to inflate its intestine with air and paddle across while afloat.

In the mammals for which water serves as a home, a source of food, or for protection, swimming may be highly developed; their body shapes and modes of locomotion can be divided into several groups.

The African otter shrews, river otters, dugongs and manatees, and whales, porpoises, and dolphins are all characterized by a more or less streamlined shape. The animals propel themselves through the water either by oscillations of the body or by movement of the body at the base of the tail, or by both. Of these, the cetaceans and sirenians have the tail highly modified as a propulsive organ, while the front limbs, which are oar-shaped, are used mainly for steering. These animals are, of course, wholly aquatic, and cannot venture onto land and survive. The otter shrews and river otters have their homes on land, but spend much of their time in the water in search of food. They move in the water by undulating the body and tail in an almost snakelike fashion. In both the otter shrews and the river otters, the tail is very powerful and is the main means of propulsion in the water. The feet of the African mainland otter shrew are not webbed, although there is a related genus in Madagascar in which they are. The front feet of otters are not, in general, more webbed than are those of their wholly terrestrial relatives in the weasel family, and in some cases they are unwebbed. Although the hind feet are webbed in some genera of otters, the feet are not excessively enlarged.

The second major group of aquatic mammals is that in which the limbs are used in propulsion. It may further be divided into several subcategories, the first being that in which both the front and hind limbs are used for propulsion. Most of the nonaquatic mammals swim like this, but the method is also used by such aquatic species as walruses, hippopotamuses, and capybaras. The second group utilizes the forelimbs as the main organs of power. This category includes the platypus, polar bear, and the sea lions. The platypus has webs that extend beyond the tips of the digits on the forefeet. When on land, this webbing is folded into the palm and out of the way. The flattened tail and webbed hind feet are used only for steering, all of the propulsion coming from the forelimbs. Sea lions and fur seals use both their front and hind feet, which are webbed flippers, for locomotion on land, but in the water they use the front flippers exclusively for propulsion, while the hind ones are trailed and aid in steering.

All of the other aquatic mammals use the hind limbs as organs of propulsion. This group includes the aquatic marsupial of South America, the yapok, which has enlarged, webbed hind feet and a long, tapered tail; the desmans, aquatic Eurasian insectivores that have laterally compressed tails and enlarged, webbed hind feet; the water shrews; the sea otter, with enlarged, webbed, hind feet, and a slightly flattened tail; the hair seals, in which the hind limbs are modified as flippers and cannot be rotated forward for land locomotion as in the sea lions, and in which the front flipper is used entirely for steering in the water and propulsion on land; and the majority of aquatic rodents (excluding the capybara) such as the nutria, water rats, muskrat, beavers, and fishing rats. All of these have webbed hind feet, usually webbed forefeet, and long tails that may or may not be modified by flattening. Beavers move in the water by alternate paddling of the hind feet, with the dorso-ventrally flattened tail used occasionally as a scull. The same is true of the muskrat, in which the tail is laterally flattened, and of the round-tailed water rat, a smaller version of the muskrat with a round tail; in both of these the hind feet are not completely webbed. The nutria of South America has partially webbed hind feet and an unflattened tail, but swims in the same fashion as the muskrat. Some mammals, such as water shrews, have fringes of bristles on the edges of the feet or tail that serve the same function as webbing or a keel.[7]

Aerial Locomotion

Although many arboreal mammals, such as tree squirrels, monkeys, and apes, may leap from branch to branch and tree to tree and thus pass a short period of time in the air completely unsupported, and flying squirrels, colugos, and some phalangers can glide, there is only one group of mammals that is capable of true, sustained flight: the bats. The wings of a bat consist of a thin membrane of skin attached to three or four greatly elongated fingers, to a lengthened arm, and to the sides (or in a few cases to the back) of the body and outer sides of the hind legs. Some bats also have a membrane that extends between the hind legs and encloses, wholly or partially, the tail. In flight, the bat's arms are moved forward and downward, then upward and backward, much like the motion employed by a human swimmer doing a "butterfly" breast stroke. Although some bats can glide for short distances, none has developed gliding to the extent of some birds which can ride long distances on upward currents of air. Some bats are capable of hovering, however, and the "hummingbird bats" utilize hovering to obtain the nectar from the flowers on which they feed.

Gliding as a means of locomotion in mammals occurs in several orders—marsupials, dermopterans, and rodents, but it generally is not the prime means of locomotion, being subordinated to walking or running. Gliding mammals all have a very similar structure, consisting mainly of a membrane of skin attached from the ankles, at least, to the sides of the body. The extreme is reached in the colugos of southeastern Asia and the Philippines in which the membrane reaches to the tips of the toes, tip of the tail, and even to the sides of the head. Colugos may glide more than two hundred feet. The flying squirrels and flying phalangers do not have as extensive a gliding membrane as the colugos, and the tail is free and is used for balance and also as a rudder during a glide. Flying phalangers may make glides of more than three hundred feet, American flying squirrels glide up to one hundred feet, and Asian and African flying squirrels make glides of a hundred to two hundred feet.

SPEED

There is relatively little information concerning the maximum speeds at which mammals can move. The speed recorded for any mammal must be equated with the terrain over which the animal passed, the distance it had to run, and other circumstances of the trial. All too often the speed has not been recorded accurately, but is merely estimated, which can easily lead to error. Among the marsupials, kangaroos are the fastest, with 25 miles an hour recorded for a chased animal; an American opossum is recorded as moving at a little over 4 miles per hour. Short-tailed shrews move at speeds up to 2 miles an hour, and on the surface of the ground an Eastern mole has been recorded at 1½ miles an hour. Among the primates, a langur has been recorded at a little more than 23 miles an hour, over a distance of 70 yards, which is a little faster than the world record for humans in the 220 yard dash, about 22.5 miles an hour. Normal leisurely walking speeds in gorillas are about a third of a mile an hour, and when chased they can move away at 5 miles an hour. It is thought that over short distances they can reach a speed of 15 or 20 miles an hour maximum.[8]

In the rodents, the fastest recorded speed seems to be about 17 miles an hour for a gray squirrel on the ground, but there are undoubtedly others not tested that can exceed this speed considerably. Kangaroo rats can travel up to 12 miles an hour, woodchucks and chipmunks 10, and deer mice at almost 7.

A European rabbit has been recorded at a speed of 20 to 25 miles an hour, and a white-tailed jackrabbit at 34 miles an hour over a distance of 50 yards. An Arctic hare is said to have traveled at 30 to 40 miles an hour when chased over a distance of 3 miles.

The carnivores, which need speed to capture their prey, are among the speedier mammals. Wolves may reach speeds of 43 miles an hour in short dashes, and coyotes 24 miles an hour. Jackals and gray foxes are reported to attain 35 and 40 miles an hour, and hyenas and lions 40 and 50 miles an hour respectively. The fastest of the carnivores, and probably the fastest of all mammals over short distances, is the cheetah, which has been clocked at 71.5 miles an hour over a distance of 700 yards.

Hoofed animals, generally preyed upon by the larger carnivores, are relatively speedy. The fastest reported is the American pronghorn, which may reach a speed of 70 miles an hour in short bursts, although its normal running speed is about 35. Wildebeest, various gazelles, and white-tail deer may reach 50 miles per hour, while zebras, wild asses, and race horses reach 40 miles an hour. A Mongolian wild ass that was chased for 16 miles averaged 30 miles an hour. The domestic horse, although reaching nearly 40 miles an hour over short distances, averages only about 11 miles an hour in longer runs. Wart hogs are reported to have reached 30 miles per hour, as have bighorn sheep, giraffes, and moose. The cumbersome black rhinoceros may reach 28 miles per hour in short bursts, and caribou can reach 25 miles an hour, as can African elephants; speeds for the last-mentioned are reported as high as 25 miles an hour for a distance of 120 yards. Racing camels, over long distances, rarely exceed 10 miles an hour.

Few speeds for bats have been recorded, but a relatively slow flier, the North American big brown bat, has been clocked as fast as 15 miles an hour. Gliding squirrels are believed to travel at a rate of about 5 miles an hour.

Many aquatic species are slow. Beavers and muskrats swim at a rate of 2 or 3 miles an hour. Right whales are slow swimmers, progressing about 2 or 3 miles an hour with 6 being a high rate. Gray whales and humpbacks are slow, averaging under 5 miles an hour at normal cruising speed, but when chased, humpbacks have reached about 10 miles an hour. Blue and fin whales are reported to be able to keep up a speed of about 24 miles an hour for 10 to 15 minutes when chased, and fin whales have been clocked at 30 miles an hour under water in short sprints. Sei whales are reported to have achieved speeds of 35 miles an hour at the surface for short distances. Sperm whales are said to go up to 24 miles an hour, while dolphins riding the bow waves of a destroyer were clocked at 38 miles an hour. Larger bottlenose dolphins have been reported to exceed 26-mile-an-hour speeds. Accurate testing at sea with small dolphins has indicated a maximum speed of 16 miles an hour over short distances. Amazon River dolphins are slower, with 10 miles an hour being near the maximum.[8]

While maximum speeds are used by animals to capture prey or to escape predators and antagonists, all animals use slower speeds in the course of their normal daily and seasonal activities, such as seeking out mates or water, or on migration.

MIGRATION

Migration is a periodic movement away from and back to a given area and should not be confused with emigration, which is a movement in one direction only, without a complementary seasonal return. Dispersal of the young is, then, a form of emigration from the initial home territory. Immigration, similarly, does not involve a return, and an emigrating mammal is also immigrating to a new area. Nomadism, a more-or-less directionless form of wandering, also should not be considered migration, because the animal does not usually return to its starting place.

Migration can be divided into several categories, depending upon the incentive to migrate. Climatic migration is a movement from one zone of climate to another. Black bears, in the northern parts of their range, may move a little to the south for the winter denning, but return to their normal summer range in spring. Cave bats may spend the summer in the attics of houses, but in fall return to caves where the temperature does not become so low as in attics and where it is certain not to drop below freezing.

Alimental migration, migration in search of food, is often directly related to climatic conditions. The great migrations of caribou in northern North America, from the tundra in the summer to the treeline to the south in the winter, is probably in search of food more than it is to escape inclement weather, but both factors undoubtedly play a part. Some whales exhibit alimental migrations, but the periodic movement from polar waters to subtropical ones is probably more related to reproduction than it is to food. Large herds of grazing animals in Africa are known to make alimental migrations in search of green grass, and the movements of the great herds of American bison probably were both alimental and climatic in nature. With the bison it was apparently a shift of some 200 to 400 miles for each of them; thus the migrants would all obtain somewhat more clement weather in the south in winter than they could expect by remaining in their summer feeding site.

All alimental and climatic migrations do not necessarily have to be over long distances. American wapiti move from the higher altitudes of the summer range to the more sheltered and less snowy valleys for the winter; at the lower altitudes they can find food more readily and also receive some protection from the winds and cold of the open mountain slopes. Local migrations are even made by house mice on farms, where the summer may be spent in the fields, close to a good source of food and where much shelter is not needed, and then in the fall the mice move back into houses and other farm buildings where there is ample shelter and a good food supply. The migrations of some bats to the south in winter is alimental, also, for some species, but much more remains to be learned about why some species migrate and others do not.

For some mammalian species, migration is concerned with

reproduction. Gray whales, which summer and feed in the Arctic, make a long migration southward to spend the winter in the lagoons of the coast of Baja California, Mexico, where the young are born and mating takes place. During the northward and southward migration little or no feeding is done by these whales. Alaskan fur seals migrate to the Pribilof Islands in the Bering Sea each summer, and it is here that the pups are born and the adults mate. In autumn the seals depart from the Pribilofs, the males wintering off the Gulf of Alaska and the females migrating some 3,000 miles southward to spend the winter off the coast of California. Although the migration to the Pribilofs is probably gametogenic in nature, reproductively inspired, the southward migration seems to be alimental.

EMIGRATION

The dispersal of the young from the parental home site is one type of emigration and is a major factor in the invasion of new areas and also in the reinvasion of areas from which a species may have been eliminated. In addition, there are species of mammals in which the permanent departure of large numbers of adults and young from the home site have been noted, and these are usually called emigrations. The lemmings of Scandinavia have been noted for their periodic emigrations during which great numbers of these small rodents have been observed moving out of their mountain tundra habitats. These "lemming years" are known to occur in cycles of approximately three years in southern Norway (and different lengths of time in other parts of the northern hemisphere) and are presumed to be a result of overpopulation, among other factors. It is known that there is a small amount of emigration among house mice while the population is low, but as it approaches the maximum carrying capacity of the environment, the number of emigrants increases.

Emigrative movements are, of course, characteristic of all dispersing mammals. However, at times the number of emigrants reaches such proportions that the animals are readily observed. In eastern North America, gray squirrels emigrating in the autumn were noted frequently in the past century, and as late as 1935 a great emigration was recorded in western New York.[9] In this case, again, overpopulation was the presumed factor that caused the great movement. Snowshoe hares, introduced Norway rats, pikas, and woodrats have been noted at times for their emigrations.

Although for most mammals the details of dispersal and emigration are lacking and their causes are still in an early stage of study, the effects of these movements are of great biological significance. First of all, dispersal by the young reduces the population in the parental environment, permitting some semblance of a stable home population. Second, if the population over a large area in the range of the species is low, the dispersing animals can repopulate the area. Third, should the natural range of the species be well populated, there will be a tendency for the dispersing animals to be forced out into areas previously not inhabited by the species. Should the newly invaded habitat lack some necessary requisite for survival, the immigrants, of course, will not live. But this will not be at the expense of the species, which will continue to occupy its existing range. Should the immigrants be able to survive, the range of the species is thereby increased. The great mortality that occurs during dispersal is, too, of biological significance. It reduces the pressure of predators on the parents, and it introduces a strong natural-selective pressure on the young dispersing animals. Those that are not speedy enough, alert enough, appropriately protectively colored, or otherwise less adapted than their siblings and fellow emigrants do not survive long enough to reproduce themselves, and thus selection for the better adapted animals takes place. These become the parents of the next generation and may introduce or maintain the genes that aided them in survival into the populations of the species.

4 ❧ Home Range, Territory, Shelter

THE dispersing mammal is leaving an established home site and is searching for an area in which it can establish itself. The natural attributes of this area to be settled, of course, vary with the requirements of each species. Damp ground may be objectionable to one species and required by another. Drinking water is a requirement for some species, and is unnecessary for others. Soft soil might be required by an animal that lives underground but is a poor digger, while a prodigious digger in the same region might find his tunnels collapsing. To some extent the selection of the home area is probably a conscious matter to the animal, but innate reaction—instinct—is probably even more important. If animals establish homes in a region that does not contain all of the requirements of their physiology, natural selection will weed them out, and the species cannot sustain a population in that locality until either the physiology of the animal has modified or the environment itself has changed.

HOME RANGE, TERRITORY, SHELTER

HOME RANGE

The area over which an animal roams in its normal activities of food getting, mating, and raising of young is called the home range. The home range of an individual may overlap or include much of the home ranges of other animals. Its size varies with the species, the size and sex of the animal, the type of locomotion, the food habits, the physiography or physical structure of the area, the time of the year, and the density of the population of the species.

Grizzly bears have been reported to have a home range about 9 or more miles in diameter, while the range of a short-tailed shrew may be less than an acre. A cottontail may have a site over 14 acres in size, and that of a raccoon may be a mile in diameter.

In general, males have a larger home range than females. In part this may be due to the larger size of males, as in the weasel family, and their consequently greater mobility. Also, males seeking mates during the breeding season would tend to wander more widely. Conversely, females raising young limit their movements so as not to be away from the den too long a period of time; as an example, the average year-round home range for the American marten has been recorded as 588 acres for the male and 172 for the female.

Jumping mammals may have a wider home range than cursorial (running) animals of similar size. The kangaroo rat (*Dipodomys merriami*) has a range of about four acres in late spring, in New Mexico, while the pocket mouse (*Perognathus penicillatus*) in the same area has a home range of from one to three acres at the same time of year. Digging rodents, such as pocket gophers, probably have smaller home ranges than ground squirrels of equivalent size living in the same area.

Carnivores tend to have greater home ranges than herbivorous mammals because they must often wander far to find their prey. Wolves and mountain lions may have home ranges covering 20 to 50 miles. A carnivorous rodent, the grasshopper mouse, may have a home range as great as 12 acres, compared with the similar-sized white-footed mouse (*P. leucopus*), a seed eater living in the same area, which ranges over only 3 or 4 acres.

The effect of physiography on home range is little studied. In the temperate zones it has been noted that meadow mice (*Microtus pennsylvanicus*) have larger home ranges in sparse grassland than they do in dense grassland.[1] This, of course, could be related to amount of food available rather than to actual physiography, as in the case of the Korean field mouse (*Apodemus agrarius*) reported to have a home range of nearly 3/4 acre in grass and shrub areas as opposed to a range of more than 1½ acres in woodland.[2]

The size of the home range also varies seasonally. Naturally, a hibernating animal is not moving about during the winter, and mammals with food stored in a burrow may not wander as widely during winter as during other times of the year. The hibernation of prey species may force carnivores to wander farther in winter than in summer in search of food. Deep snow, conversely, may impede movement and confine large ungulates such as deer to the relatively small confines of a "deer yard," an area where the animals have packed down the snow.

During the breeding season males may travel farther in search of mates, and at the times females are raising young their range may be smaller than at other times of the year. Some indication of this has come from a study of kangaroo rats in New Mexico where it was noted that in the month of

March, when newborn young were in the burrows, a female had a home range of about 1/3 acre. In April, when some of the young were coming out of the burrows, the home range of the same individual was about 2¼ acres, and in May the home range averaged 3⅓ acres.[3]

The home range of an individual may overlap that of another animal of the same species. Thus, in a single acre, there might be three individuals each with a home range of 3/4 acre. Mammals do not defend their home range, but rather share it with others of either sex of the same species (and even of other species).

TERRITORY

Animals with a fixed home that, naturally, lies within the home range, generally maintain an area around the homesite that they defend against other individuals of the same species. This defended area is called the territory. While the phenomenon of territoriality is well known for breeding birds, it is much less understood for mammals. Many small rodents probably have territories, but because they are also nocturnal, direct observation of territorial defense is rarely recorded. A further difficulty in establishing territoriality on the part of some mammals is the occurrence of a social hierarchy (see Chapter 7). When a "peck order" exists in a group of mammals, the reactions of a dominant individual toward a subdominant one may give the appearance of territory defense, even though no physical territory exists.

One of the most dramatic examples of territoriality is shown by some of the eared seals. As noted earlier, the adult male Alaskan fur seals come ashore on the Pribilof Islands of the Bering Sea in late April or May, and they establish their terri-

BIOLOGY OF MAMMALS

tories on the beach. The boundaries of these territories, 75 to 100 feet in diameter, are best known to the owner, but are quickly learned by intruders who face a savage attack by the bellowing, 600-pound, sharp-toothed beach master. Bulls under six or seven years of age studiously avoid the established territories and congregate by themselves in bachelor herds. The first bulls to arrive obtain the choice sites, nearest the water. Later arrivals, though mature and strong, must take territories back from the water's edge, and are called "idle bulls." They have much less of a chance to obtain mates, for with the arrival of the females from mid-June to mid-July, the bulls herd their mates into their territories and fight off all attempts by other males to steal cows. From May to August, the life of the territorial beach masters is one of constant vigilance to protect their territories and the cows within them. During the period they do not leave their territory nor do they eat. By August, when the virgin females arrive, the beach masters are so thin and worn from months of vigilance and breeding that most of the new cows are bred by the idle bulls in what had been less advantageous territories.

Various degrees of territoriality have been noted for tassel-eared squirrels, red squirrels, eastern and western chipmunks, chickarees, Mexican ground squirrels, fox squirrels, European rabbits, and pikas, as well as most of the eared seals. As noted before, all of these animals are active during the day.

While the territory is protected, the defense rarely comes to the point of physical contact between the trespasser and the defender, although probably among fur seals it is more common than among most other territorial mammals.

The territories of clans of howler monkeys are marked by vocalization. When two clans come near each other at the edges of their territories, there ensues a vocal battle, mainly by the males, but at times supplemented by the barks of females and whines of young. These vocal battles may continue for

HOME RANGE, TERRITORY, SHELTER

several hours if one or the other group does not remove itself. Every dawn the males of each clan also howl. This has been interpreted as a means of letting the other clans know the approximate location of the group, thus indicating the territory.

Many mammals probably mark their territory or home range by means of scents. The use of olfactory mechanisms in communication is one of the least studied facets of mammalian life history, but probably is one of the most important aspects of the life of mammals. Glandular secretions from the legs, hoofs, and heads of ungulates may play an important part in the social life and reactions of these animals. Territorial scent marking has been reported for American wapiti, American badgers, hippopotamuses, European bison, and dogs and probably is common in most members of the weasel, dog, and civet families. Anyone who has watched male dogs sniffing and then marking hydrants, automobiles, and building walls on a city street will readily recognize the extent of olfactory communication in these animals.

Visual territory marking probably also exists for some mammals. Small herds of wildebeest in Tanzania are reported to be led by a male who marks the center of the territory by a bare patch of ground, a "stamp." Intruders are subject to a ritualized display of headshaking and mock charges by the defender of the territory; persistent transgression of the small territory, usually less than two hundred yards in diameter, may result in physical contact in the form of butting. However, injuries from territory defense seem to be rare.

SHELTER

Mammals maintaining home ranges and territories have their homes within these areas of familiarity or defense. If home be taken to mean a shelter, then many mammals are homeless. Musk-oxen, for example, as well as many other hoofed animals, may pass their entire lives virtually unsheltered from the elements. The same is true of many kinds of whales, which are not known to seek shelter, other than the quiet waters of the depths, at any time of their lives.

The first step toward a home is in the utilization of some natural structure for shelter. Desert bighorn sheep seek a cave or rock shadow to avoid the extremes of midday summer heat. Seals may go to the quieter waters of a cove during a storm, and California gray whales return thousands of miles each year to the calm, shallow lagoons of Baja California to bear their young. African antelopes frequently cluster under a tree to avoid the sun during the hottest part of the day, and American jack rabbits shelter under clumps of cactus brush. Murine opossums may sleep curled in a dead banana leaf. A ravine or gully may provide temporary shelter for an otherwise homeless hoofed animal during a storm—although the females of these species usually seek some form of natural shelter for seclusion at the time they drop their young.

A more advanced type of home involves some alteration of the environment by the animal. The restless pacing of hoofed animals at term softens the ground where the young will fall, even though this small area will serve as a home for only a few hours or days. Some tropical bats (*Uroderma bilobatum*) cut palm leaves in a fashion that causes them to droop in a tentlike form and thus provide the darkness and shelter that the bats desire. Female cottontails excavate a shallow depression in the ground, line it with their own belly fur, and here deposit their young.

Although many monkeys, especially the South American ones, seem not to have homes and spend their nights sleeping on the branches of trees, gorillas and chimpanzees make nightly beds of bent vegetation low in trees or on the ground, and use them only once.

A more permanent home often consists of a burrow in the ground, although relatively few kinds of mammals lead wholly subterranean lives. North American and Eurasian moles, African golden moles, and the Australian marsupial mole are the most subterranean of all mammals. The tunnels of American moles are generally of two types: a deep burrow some two feet below the surface used as a nest chamber, and a surrounding network of shallow tunnels that are the paths that the moles use to search for food. It is the latter that break the surface of the ground to disfigure lawns and golf courses, and these ridges have been traced for half a mile, although it is not known whether or not they are the work of only one mole. Dirt from the excavations is pushed out onto the surface as molehills. Townsend's mole of the western United States is known to have made 302 mounds in a 1/4 acre field in 77 days.[4] European moles make similar burrows and molehills and line their nest chamber with dead leaves and dry grass, as do the American moles.

The American pocket gophers are also prodigious burrowers, but they have a way of life different from moles: They are rodents and vegetarians, and they do some feeding on the surface of the ground. Their burrows go far deeper and are longer and larger than those of moles. A pocket gopher in sandy soil may make two to three hundred feet of tunnel a night. Its burrow is elaborate, with underground storage rooms, sleeping quarters, side tunnels, and even a separate chamber for a toilet. In South America, the tuco tucos live a similar existence.

Many other rodents are burrowers, and the structure of their excavations varies by species and soil type. The value of a narrow entrance, dark twisting passages familiar only to the owner, and the shelter from climatic changes is obvious, but some species add further devices for their protection. The danger to a subterranean home from flooding is great, and European moles make a drain at the base of the cavity that they use as their nest. Prairie dogs make a mound as much as two feet high around the opening of the burrow that protects the entrance from light floods. In the burrow system itself, the first lateral tunnels ascend slightly, thus leaving the chambers at the ends of these branches safe from water that might flow over the protective mound at the entrance and penetrate the burrow proper.

Mounds at the burrow entrance are also characteristic of kangaroo rats, inhabitants of the western American deserts. But flooding is probably of minor consequence to these animals, for they plug the entrance holes with dirt during the times that they are in the burrow. The dirt plug not only discourages predators such as snakes, but serves a more important function by protecting the kangaroo rat from the rigors of desert climate: low humidity and extremely high daytime temperatures. By coming out of their burrows only at night, the kangaroo rats are active when the temperature is much cooler and the humidity is higher. By plugging their burrows in the early morning hours, the animals maintain a climate that never becomes hotter than about 93°F. (and generally is around 10° cooler than this even in the summer). For an Asian mammal of similar habits, a gerbil, it has been shown that the burrow temperature 4 inches inside the entrance was only 81°F. when the outside surface temperature of the ground approached 140°.

Even animals that make much shallower burrows in the ground derive protection from predators and from weather. The brown lemming of Alaska may include in its summer burrow a chamber that descends into the permanently frozen soil, the permafrost, thus providing the animal with a cool retreat that, because it is in frozen ground, also provides ice-solid walls that would be hard for a predator to dig through. During the winter, lemmings must make their nests above the frozen ground and beneath the snow. The insulating effect of snow is so great (as the Eskimos well know) that 20 inches beneath the surface the temperature may be 70° higher than the air temperature.

Other forms of shelter, often less modified by the animal, include caves, rock crevices, and hollow trees. Even these are altered slightly to suit the animal. Raccoons drag leaves into a crevice to provide insulation or softness. Eastern fox squirrels continuously gnaw back the wood at the entrance to a tree-hole den to keep the wood from growing back and reducing the size of the hole.

Grass and leaves make up the nests of many tree and ground dwellers. Squirrels make leaf nests using an outer layer of twigs with leaves attached, an inner layer of damp leaves pressed together, and a soft lining of shredded bark and leaves. A fox squirrel may use one of these nests for several years.

Muskrats and beavers are noted for their house building. Although both of these rodents may live in burrows on the banks of rivers or streams, they frequently construct homes of considerable size. Muskrat houses are made of aquatic vegetation—cattails, sedges, grasses—originally water-soaked and piled in the water. While the mass settles, the muskrat makes an underwater channel to it and through the center. As the vegetation is piled up, a single room above the water is made by pulling away the central plants; it has alcoves to provide separate sleeping and eating quarters. There may be two escape routes, one at each end of the chamber. Most muskrat houses are in water about a foot deep; they are roughly conical and about the size of a bushel basket. Rarely, they may be as large as ten feet in diameter and four feet high. Although they are not excessively strong, the houses provide the muskrats with protection. During the summer, if a predator starts to tear away the structure, the muskrat can escape through an underwater exit. In winter, the moist vegetation composing the nest freezes, making it quite difficult for a predator to tear open, even though the attacker has no difficulty reaching it by crossing on ice.

A beaver lodge is usually located in water about four or five feet deep. It is constructed of twigs, branches, and small tree trunks plastered with mud and stones and may reach a height of seven feet and a diameter of thirty feet. The large single room is as much as four feet in diameter and three feet high, and its exit leads out underwater to the pond. The thick walls of mud and wood, hard baked in summer and frozen in winter, provide excellent protection for the beavers.

Most other mammals generally limit their alteration of the environment to the small area of the nest, but the beaver makes a major change; the construction of its dam produces a rise in the water level of the stream, thus providing deeper water to surround the lodge and also for escape by diving. Because of its depth, the beaver pond does not freeze to the bottom, allowing the animals access to the food they have stored there. One of the results of such a permanent home and alteration of the environment is a transportation problem, for the available food supply is at an increasingly greater distance the longer the lodge is occupied. So, beavers also build a communications network of canals by which they can reach new sources of food and float it back to the home without sacrificing the protection of water. Huge dams and houses and lengthy canal systems are the works of generations of beavers.

Only one other mammal alters his environment to a greater extent than the beaver—man. In what little is known of human history, we probably have an idea of the whole course of mammalian homes and shelters. From tree-dwelling with, at best, an apelike temporary bed, to a terrestrial existence

HOME RANGE, TERRITORY, SHELTER

in caves, to an increased alteration of the environment through crop raising, housebuilding, land leveling, dam building, marsh draining, and control of other animals and plants, man has produced environmental changes that threaten the existence not only of all other mammals, but of man himself.

5 ❦ Air, Water, Food

ALTHOUGH a home or shelter may not be needed by some mammals, all of them have certain basic physiological requirements in order to stay alive. In the hierarchy of necessities for survival, oxygen, water, and food are essential, in that order, to the lives of mammals. They can survive for the shortest time when deprived of oxygen, somewhat longer without water, and for a considerable time without food. Using man as an example of a mammal that does not live in water, underground, or in some other specialized environment, permanent damage occurs to brain tissue if it is deprived of oxygen for as little as four minutes. Without water a man may not survive a single day if he is exposed to desert heat and sun, and only about four days in a temperate climate. Without food and with moderate exercise, an adult human in good condition can easily live for a week; under circumstances of idleness, he can survive three to four weeks without permanent damage.

AIR, WATER, FOOD

AIR

For most land mammals, deprivation of oxygen as a cause of death would occur through drowning, or more rarely, through the collapse of a burrow. Just what the oxygen requirements of such underground dwellers as moles and pocket gophers are has not been investigated, although one might suspect that they might have a greater tolerance to lack of oxygen than other mammals. For aquatic mammals, of course, a return to the surface for breathing is essential, but how long most of them can survive between breaths is not well known. Platypuses can remain under water for 5 to 10 minutes, and manatees from 7 to 16 minutes. Sea otters have survived underwater for 5 minutes; polar bears, 1½; and hippopotamuses, 15 minutes. Elephant seals have survived without surfacing for about 7 minutes, muskrats for 12, and seals and beavers for 15 minutes. Whales have been recorded as needing air within well under a half hour, but rorquals (*Balaenoptera borealis*) and blue whales are known to have submerged for about 50 minutes, humpbacked whales for an hour, sperm whales for as much as 1¼ hours, a bowhead whale for 1 hour and 20 minutes, and a bottlenose whale for 2 hours.[1]

Young animals are often much more tolerant of oxygen deprivation than are adults. A newly hatched platypus can survive 3½ hours of submergence and newborn Norway rats can survive 2 hours in an atmosphere of 100 per cent nitrogen.[2]

WATER

While a mammal needs water, it does not necessarily need drinking water. As more and more studies are made, it is being discovered that many mammals obtain their water from succulent vegetation or from the metabolic breakdown of food and may pass many months, or even their entire lives, without actually drinking water. Mountain gorillas, living in the humid forests of central Africa and feeding on juicy foods, are reported not to drink in the wild, although they do in captivity.[3] It has been the mammals of the hot, dry deserts, however, that have attracted the attention of physiologists studying water metabolism. Mammals living in these climates have generally adapted to the heat and lack of water either through seeking a niche that does not expose them to the extremes of the climate or through physiological adaptations, or both.

A mammal loses water primarily through the urine and feces, from moisture in expired air from the lungs, or by sweating. By their adapted physiology, kangaroo rats have reduced their water loss to a minimum. The urine of these animals is almost four times as concentrated as that of man, and a kangaroo rat can thus excrete proportionately more waste with less loss of water. The feces of these animals are very dry, which also helps water retention. The rate of water loss from the lungs is extremely low, and because of the humid environment of this rat's plugged burrow, it is reduced to a minimum.

In addition to their physiological reduction of water loss, these desert mammals are able to exist on their diet of dry seeds, and they need never drink. By the metabolic breakdown of the seeds, carbon dioxide and water are produced and kangaroo rats are able to utilize this metabolically produced water to supply all of their water needs. Although all mammals produce carbon dioxide and water from the breakdown of carbohydrates, it is only desert animals that are known to have their water balance adapted to the point where these by-products are of great use to them.

Other desert rodents that cannot utilize dry food to produce energy and metabolic water must eat succulent vegetation to obtain moisture. Among their adaptations to this way of life are estivation and hibernation during the times of year when green food is not available, plugging of burrows, and enlarged body surfaces (ears, feet, tails) through which heat may be lost from the body without necessitating sweating.

The adaptations of one-humped camels, after being speculated upon for many years, have finally been studied. Although they cannot utilize burrows like kangaroo rats, camels receive some insulation from their curly wool; in addition, they concentrate their urine. A camel's body temperature may drop as low as 93°F. during the night, and the animal does not begin to sweat until its body temperature exceeds 105°, thus the camel can tolerate a 12° rise in body temperature before it begins to lose water. Moreover, it can survive a water loss of about 24 per cent of its body weight without serious effect; man dies when his water loss reaches 12 per cent. All of these physiological factors aid in desert survival. If green food is available and the temperatures not extreme, as in winter, camels may go months without actually drinking water and still remain in good condition. When dehydrated, they can make up their water loss rapidly—one camel reportedly drinking 27 gallons of water in 10 minutes.[4]

Strange as it may seem, whales, porpoises, seals, sea lions and other marine mammals may also have a water problem. Surrounded by a salty medium and with generally no access to fresh drinking water, these animals have body fluids that are less salty than the sea. Thus, whenever the water of the ocean and the body fluids are separated by a semi-permeable membrane, there is a flow of the less-concentrated water from the animal into salt water. This, if unchecked, would eventually result in the internal environment of the mammal becoming as concentrated as the sea. This dehydration, however, does not happen in nature.

Just how these marine mammals manage their water metabolism is not known. Very probably, they ingest as little sea water as possible. The food of the toothed whales and most of the seals consists of fish and other vertebrates that do not have a salinity greater than their own, and this would tend to minimize the animal's need for water. In addition, as noted, the breakdown of food liberates water, and some of this may at least partially supply the required amount for survival. It is believed that whales must produce large quantities of urine to remove salts from their bodies and that metabolically produced water is used for this purpose.[5] In addition, whales have no sweat glands and do not lose moisture in that fashion. The air they breathe is nearly saturated with water, and so there should be little water loss through the lungs. The problem of the baleen whales, however, seems to be greater than that of the toothed whales. Feeding on invertebrates with a salinity about the same as sea water, baleen whales are ingesting large amounts of salts, which have to be excreted in some fashion, either through great quantities of dilute urine or smaller amounts of concentrated urine. Which mechanism is used and how it is accomplished remain to be discovered.

Seals have a similar problem, and bull Alaskan fur seals, at least, do not drink during the two months they are ashore, nor has any of them been observed drinking sea water. However, in captivity, they customarily drink fresh water, though there is a report of a captive Antarctic leopard seal drinking sea water.[6] With the increasing knowledge of whales, seals, and walruses being gained from animals in captivity, it is hoped that some answers to the problems of water metabolism in marine animals will be found.

For most other mammals drinking water is essential if

insufficient moisture is obtained from the food. In addition to the moisture in the plants and animals that are consumed, water may be obtained from dew on leaves and grass, melting snow, water trapped in plants and hollows, and from ponds, streams, and lakes.

FOOD

For the energy to carry out all of their varied life processes, mammals must have food. The three basic kinds are carbohydrates, fats, and proteins. The source of carbohydrates—starches and sugars—is from plants, while fats and proteins may be obtained both from plants and animals.

Each type of food has a different amount of energy available: carbohydrates have about 4.1 Calories per gram; proteins have 5.7 Calories; and fats are the richest in caloric content, with 9.5 Calories in each gram. These values, however, are those that would be obtained from the foods if they were fully oxidized; the actual value to an animal may be less. A pound of protein, for example, has a caloric value of 2,563 Calories, but this food is never completely oxidized by mammals and part is wasted. The actual energy value to a mammal of a pound of protein may be only about 1,973 Calories. The remainder is excreted in body wastes, reducing the actual value of protein to 4.4 per gram, close to that of carbohydrates.

The portion of a weight of food available for digestion by an animal is far less than its caloric value, and, as stated above, of the digestible amount, less still can actually be assimilated and absorbed. A hoofed animal being fed on a dry food with a caloric value of 4.5 Calories per gram actually needs about twice the weight of food theoretically necessary to obtain its minimal energy requirement. When feeding on moist food, the number of Calories per gram would be even lower, because of the weight of the water, and thus an even greater quantity of food would be required.

Although mammals eat a great variety of foods, all are broken down into simpler components when digested so that they can be absorbed by the body. Carbohydrates are acted upon in the mouth by secretions from the salivary glands, and starches are reduced to sugars. In the stomach, glands secrete hydrochloric acid, and this, plus enzymes such as pepsin and rennin, works on other foods. Pepsin breaks proteins down into peptones and proteoses; rennin coagulates the casein in milk; gastric lipase works on emulsified fats. In the small intestine, erepsin, another enzyme, breaks proteoses and peptones into amino acids; maltase breaks the sugar, maltose, into glucose; sucrase breaks another sugar, sucrose, into glucose and fructose; lactase converts lactose into glucose and galactose. Trypsin helps break down proteins into amino acids; lipase converts fats into fatty acids and glycerol. There are numerous types of enzymes that function in the reduction of foods into simpler components so that they can be absorbed through the walls of the intestine.

Once foods are absorbed, they go to different parts of the body. In mammals, the sugars are transported to the liver and are converted into glycogen (animal starch), and either stored or released for energy into the blood stream as the sugar, glucose, which is the main source of energy for muscular work and heat production. Proteins are also carried to the liver where they become available for use in cell construction and body repair. The nitrogenous parts of protein metabolism are passed from the body as waste in the urine. Fats are carried through the lymphatic system and then to the blood stream.

They are afterward converted into animal fat and stored for later use. Excess carbohydrates also can be converted to fats as a reserve energy supply for the animal.

Plant Foods

As the primary source of food for all animals, plants are consumed by the majority of mammals, both in numbers and kinds. Most of the rodents, the largest order of mammals, all of the lagomorphs, and most of the hoofed mammals are primarily vegetarian in their food habits. Although nearly all parts of plants are consumed, it is doubtful that any mammals are capable of consuming every element of all plants. Bark is eaten by porcupines and beavers, as well as by some kinds of mice and rabbits, and by elephants and many kinds of deer. Wood itself is rarely sought as food, but browsers, such as moose, deer, and black rhinoceroses, consume much of it in their twig-clipping food habits. Leaves and grasses are widely eaten, especially by rodents such as muskrats, prairie dogs, Australian possums, phalangers, and by most of the kinds of hoofed animals.

The fruits, berries, flowers, nectar, and seeds of plants are also eaten, of course. Not only do the more or less strict vegetarians feed on them, but many of the omnivores—North American opossum, foxes, bears, skunks, and mongooses—utilize ripe fruits and berries as a major part of their diets. In addition to the many primates that are frugivorous, there are bats that are fruit eaters, as well as some tropical members of the Order Carnivora, such as the kinkajou and olingo.

Although various mammals, like some Australian possums, feed on flowers, relatively few have specialized for a diet largely or exclusively of nectar. Several kinds of American bats, however, do feed on nectar, as does the marsupial honey "mouse" of Australia.

Seeds, of course, are utilized by a great many mammals, especially rodents: tree squirrels, although they are not exclusively seed eaters, and the desert-dwelling kangaroo rats, pocket mice, and gerbils. Even bears and deer feed on acorns, and nuts are a staple in some primate diets.

Roots and tubers are generally the foods of the more subterranean mammals, such as pine mice and pocket gophers. Peccaries and wild pigs root up bulbs and tubers as part of their omnivorous diets. Fungi, both above and below the ground, are utilized for food by mammals. American red squirrels, especially, eat mushrooms, including the kinds poisonous to man, and the same has been reported for the European red squirrel and even for the roe deer. Lichens make up much of the diet of caribou in both winter and summer.

Animal Foods

INVERTEBRATES

In both numbers and kinds, invertebrates are far more plentiful than vertebrates and thus are an excellent source of food. Ants and termites, perhaps because of their colonial habits, have attracted mammals as predators, and members of several orders have become specialized for feeding on them. In South America there are three kinds of anteaters: the giant and silky anteaters and the tamandua. In Africa and Asia the pangolins specialize in feeding on termites, and also in Africa the aardvark and aardwolf both consume ants and termites. In Australia and New Guinea the monotreme echidnas are specialized anteaters, and, somewhat less specialized,

AIR, WATER, FOOD

the marsupial numbat preys on termites in rotting logs. Although all of these animals are called anteaters, their diets are probably composed more of termites than of ants.

Other insects are eaten by many mammals. Some of the larger carnivores, such as wolves, coyotes, and even bears, eat bees, grasshoppers, and crickets. Many kinds of bats are almost exclusively insect eaters, catching their nocturnal flying prey on the wing. Most kinds of mice and rats occasionally eat some insects, and they are the primary food of moles, shrews, skunks, and many primates at various times of the year.

Crustaceans, in fresh or salt water, are eaten by many mammals. Raccoons, mink, and otters prey on fresh-water crayfish. On coasts, crabs are eaten by raccoons, coatis, and even foxes and dogs. In open seas, euphausiid shrimp make up the food of most of the baleen whales. Other invertebrates and small vertebrates are undoubtedly ingested as the whales engulf enormous mouthfuls of this floating "krill," as the euphausiids are called. The fish-eating bats of the Gulf of California also eat crustaceans.

Mollusks, too, are an important food for some mammals. The sea otter feeds on them, especially the abalones of the California coast, and the diet of the walrus is almost exclusively mollusks. Squid are a major source of food for the sperm whales and southern elephant seals, and the sea lions of the Falkland Islands also are reported to eat cuttlefish. Omnivores such as skunks, and badgers, shrews, and moles eat land snails.

Other invertebrates and invertebrate products are eaten by mammals as they can be obtained. Moles and shrews eat earthworms, and the platypus is also reported to eat worms. Echinoderms, starfish and sea urchins, are eaten by walruses and sea otters, respectively. Honey is a favorite food of bears and ratels.

BIOLOGY OF MAMMALS

VERTEBRATES

Fish are, of course, important in the diets of some aquatic mammals, such as the fresh-water otters; brown bears are notorious as salmon catchers, feeding on those that have returned to the rivers to spawn. Most carnivores will not turn down fish if they can catch them. In India, a small cat feeds on them so frequently that it is called the fishing cat. Many seals and sea lions feed mainly on fish, as do most kinds of dolphins and porpoises. Among the more unusual fish eaters are at least two kinds of bats of the American tropics and subtropics that feed on small prey snagged from the surface of the water with their elongated, curved hind claws.

Reptiles and amphibians are a part of the diet of a large number of the smaller and medium-sized carnivores and omnivores. Lizards and snakes are eaten by badgers, foxes, skunks, pigs and peccaries. Mongooses have a reputation for snake killing, although they do not confine their diet to these animals. Frogs and toads are eaten by raccoons, mink, and otters. Adult turtles are sufficiently well protected to survive attacks from most carnivores, but their eggs and young are relished by skunks, raccoons, coatis, opossums, coyotes, and any other omnivore that can obtain them.

Birds are eaten, when they can be caught, by most of the carnivorous mammals and many of the omnivores. Even in flight a bird may not be safe from certain agile primates; gibbons have been seen to leap from tree to tree, snatching them while they are in flight. Ground-nesting birds are eaten by wild cats and dogs, weasels, mongooses, and civets, and their eggs are equally vulnerable.

Mammals are probably the main source of food for those mammals that feed on vertebrates. Almost every herbivorous species is subject to predation from omnivores and carnivores; even a medium-sized carnivore, such as a 40-pound wolverine, is capable of attacking and killing a snow-bogged moose that may be twenty or more times its weight. Even smaller carnivores or omnivores are not safe from predation by the larger ones. In Siberia, bears are killed by tigers, and killer whales seem to prefer seals as food.

Rodents, lagomorphs, and hoofed animals, however, are the prime source of food for mammalian carnivores. How important the presence of carnivores is in controlling the numbers of these herbivores has been shown many times in the overpopulations of deer, mice, and rabbits that have occurred when mountain lions, wolves, and foxes have been eliminated from an area. The overabundance of herbivores is then usually followed by a destruction of the vegetation, starvation for the herbivores, and erosion, flooding, and general alteration of the physical and biological characteristics of the region.

Many carnivores and omnivores will feed on carrion at times, and a few prefer scavenging. The hyenas of Africa and Asia are capable of hunting and killing for themselves, but may not commonly do so, choosing instead to feed on the leavings of the other carnivores. Jackals, too, have some tendencies in this direction.

One other source of food, unique among terrestrial vertebrates, is blood. Three genera of American bats live exclusively on a diet of blood. These vampires now obtain their food from domestic mammals, poultry, and man where available. There seems to be some preferences among the three vampires, with one, at least, apparently preferring the blood of birds. The amount of blood taken in a single feeding by a bat is insufficient to kill most of the hosts, and this sanguinivorous habit is essentially a form of parasitism.

TYPES OF FEEDERS

Herbivores

The tough cellulose walls of plant cells are not only difficult to crush and grind, but are also indigestible to mammals. The consumption of any parts of plants requires special adaptations by the mammals that utilize them. For gnawing and snipping, many plant eaters have sharp and strong incisor teeth. In the rodents, especially, they are well developed. Rodent incisors continue to grow throughout the life of the animal and must be worn down, or the mouth would be forced open so that the grinding teeth behind would not occlude properly. The animal works the upper and lower incisors against each other. In the process, the harder enamel on the front surface of the teeth is worn away less easily than the softer dentine behind, and a beveled, chisel edge is thus obtained. In hoofed animals the incisor teeth also perform a snipping function, although such herbivores as deer and cattle have only lower incisors, and these, together with an incisiform canine tooth, work against a horny upper pad for tearing or snipping. Many herbivores lack canine teeth; this is typical of rodents, rabbits, and many ungulates. However, some fruit eaters, such as kinkajous and fruit bats, have well-developed canines that are probably used to hold the food and also to cut through its skin.

The cheek teeth of herbivores are designed for crushing foods or for grinding them. These molars and premolars are usually fairly broad and long, having a large working surface. In horses and cows and similar herbivores, the cheek teeth are high (hypsodont) and have a complex pattern of enamel, dentine, and cement. As the tooth is worn by abrasion, the different hardnesses of its materials leave a rough grinding surface.

The normal life span of an animal is governed to some extent by the life of its teeth, and herbivores, feeding on coarse roughage, have adapted to extend the life of their molariform teeth in several ways. The high crowns already mentioned are one method. Some rodents such as capybaras have, in addition to the incisors, molars that persistently grow throughout their lives. Another method to extend the length of usefulness of the teeth has been achieved by the elephants. In their comparatively small mouths, elephants have room for only one huge, ridged molar in each side of each jaw. As these teeth are worn away, they are replaced by a new tooth growing in from the rear, until the life total of three molars is exhausted.

In addition to the great wear that vegetation causes on teeth, herbivores face an even more serious problem. All plant eaters require the assistance of bacteria, yeasts, or protozoa in their digestive tracts in order to break down the undigestible plant cellulose into starches and sugars that can be absorbed. As time is required for the microorganisms to act, the digestive tract of herbivores is much longer than those of carnivores. The caecum, the pouch in which some bacterial digestion of plants takes place in nonruminants, may occupy as much as 52 per cent of the total length of the intestine in the muskrat, 37 per cent in the cottontail, and 8 per cent in the guinea pig. The large intestine of herbivores also occupies a much greater percentage of the total length of the digestive tract than in the carnivores. Woodrats have 47 per cent of the total intestinal length occupied by the large intestine alone and a caecum that takes up 7 per cent. An elephant, which lacks a caecum, has about a third of its digestive tract composed of the large intestine. In contrast, omnivores have only 3 or 4 per cent of the intestinal length taken up by the caecum and only about 15 per cent by the large intestine. Carnivores generally lack a caecum, or have one that is only 1 or 2 per cent of the intestinal length, and the large intestine is usually about 10 per cent or less the total.

One large group of hoofed animals, the ruminants, has evolved another complex digestive mechanism to handle plant foods. Cattle, sheep, goats, deer, camels, and antelopes are all ruminants. These grazing or browsing animals chew their fodder slightly and swallow it. It passes down the esophagus into a large chamber, the rumen, which is actually a pouch of the esophagus and not a development of the stomach. There microorganisms, especially bacteria, start the breakdown of the plants. In the second pouch of the digestive system, the reticulum, additional bacterial action takes place. Grain and seeds pass directly to the reticulum after being chewed and mixed with saliva. From the rumen and reticulum, the food is regurgitated to the mouth, where it is thoroughly chewed once more and reswallowed. Instead of going to the rumen, the food now passes to the third esophageal chamber, the omasum, where much of the moisture is squeezed from it; and from the omasum the food is passed to the stomach, the abomasum, where the bacteria and other microorganisms are killed by the acid secreted there.

In addition to breakdown of cellulose, these bacteria themselves serve as food for the ruminant. Their metabolic action in digesting vegetation produces other nutritive materials. In the rumen of a sheep at the height of digestion there may be thirty times as much fatty acid present as there was in the ingested food, and the fatty acids can be utilized as food.

Although the protozoa present in the rumen serve useful purposes, they are not essential, and animals in which they have been killed suffer no nutritional deficiency. However, if the bacteria in the digestive system are eliminated, the animal may suffer malnutrition or starve, even though ample food is present. Not only do the bacteria break down the plants to release the carbohydrates, fats, and proteins in the plant cells, they produce the fatty acids already mentioned and synthesize the necessary B-complex vitamins and all of the essential amino acids.

Other mammals metabolize plants by bacterial action, in general prolonging the time food is in the digestive tract in order to give bacteria sufficient time to complete the breakdown. Some mice and rabbits practice coprophagy, the reingestion of fecal pellets. These pellets are passed from the caecum, and the animal eats them as they come out of the anus. They contain large amounts of vitamin B_{12} and a vitamin deficiency may occur if the animal is deprived of them.

Although most herbivores eat a variety of foods, a few are very limited in their selection. The three-toed sloth feeds almost entirely on the leaves and flowers of the Cecropia tree; little is known of its digestive processes. The Australian koala lives on the leaves of about a dozen kinds of eucalyptus trees, and no others, while to most mammals the oils in these leaves are poisonous; even the koala apparently must change its food source at certain times of the year. When baby koalas are being weaned, the mother feeds them a fecal pap from her cloaca. It may well be that, in addition to nutritional value, this pap is supplying the young animal with the proper digestive bacteria so that it can later feed on the eucalyptus. The red spruce mouse of Oregon and northern California feeds exclusively on the needles of the Douglas fir.

Aside from their internal adaptations to feeding on plants, many herbivores have external specializations. The prehensile lips of black rhinoceroses, the flexible proboscises of elephants and tapirs, and the long, flexible tongues of giraffes and okapis all aid in reaching food. Prehensile-tailed South American monkeys can hang from a limb by this appendage to reach food on a lower limb, and the two members of the Order Carnivora that have prehensile tails, the kinkajou and the binturong, are both arboreal frugivores. It is more likely that this adaptation was in response to a life in the trees rather than to an herbivorous diet. Forepaws capable of manipulating objects are common to most monkeys and many rodents, especially nut eaters. Nectar-feeding bats have very long, slender, hairy tongues to obtain their food. The long hoofs of the caribou are used to paw away snow to uncover lichens in winter.

Carnivores

Carnivorous mammals seem to have required less digestive specialization than the herbivores. The teeth, although specialized in some, generally are of a more primitive type, perhaps because the earliest mammals were flesh-eaters themselves. The incisor teeth of carnivores are generally chisel-edged for snipping and, unlike those of rodents, do not grow throughout the life of the mammal. The canines are usually well developed for stabbing, holding, or slashing. The premolars are relatively simple, pointed teeth to assist in cutting and holding food. The same is more or less true of the molars, except that in the most carnivorous mammals, members of the Order Carnivora, a pair of teeth on each side are modified for shearing. These carnassial teeth are best developed in weasels, dogs, and cats. The other molars may be reduced in number or somewhat flattened for crushing and grinding, especially in the more omnivorous mammals, such as bears.

There are no especial adaptations of the intestinal tract, except that it is short in comparison with that of the plant eaters, and most carnivorous species lack the caecum so evident in herbivores. In order to catch their food, carnivorous mammals generally need a higher degree of agility and intelligence than the herbivores. They often have a very fast reaction time. They usually have good eyesight, a highly developed sense of smell, or some other mechanism for the detection of prey.

Carnivores often kill by using their canine teeth for stabbing, their sharp claws for holding and tearing, or even just their great strength in breaking the neck of their prey. Most of the carnivores that feed on mammals, birds, and reptiles have made similar adaptations for eating, and many are not specialized eaters, feeding on anything they can catch.

Insectivorous mammals, however, show a few specializations specific to their diet. The teeth, as in some insectivorous bats, do not have the upper-lower shearing molar-premolar combinations of some of the vertebrate predators. Instead, their cusps form a W-shaped pattern, with the cutting and crushing action being performed as the upper and lower W-shaped teeth occlude. Shrews—which feed on many kinds of insects, worms, other invertebrates and even small mammals—have such teeth.

In addition, some shrews have a salivary gland that contains a venom capable of killing as large a victim as a mouse. Shrews are the only mammals known to use a venom to obtain food, although male platypuses have a functional poisonous gland that can inject poison into an enemy through a hollow spur on the hind leg. Bats, dolphins, and probably some shrews locate their prey by means of echo-location—the emission of shrieks, often high-pitched, and hearing the echoes that bounce back. The aye-aye, a lemuroid of Madagascar, has a greatly elongated middle digit on the forefoot, tipped with a long claw. It uses this to pull out insect larvae and wood-boring beetles from chambers in the wood of trees. A similar adaptation has been acquired by the striped possums, marsupials of New Guinea.

An extreme of specialization for insectivorous life is found among the ant and termite eaters. Anteaters from different orders of mammals are found in various parts of the world. The echidnas of Australia, Tasmania, and New Guinea are monotremes; the numbat of Australia is a marsupial; the giant anteater, the tamandua, and the silky anteater of South America are all edentates. In Africa and Asia the pangolins feed mainly on termites and belong to the Order Pholidota. Also in Africa are the aardvark, the only living representative of the Order Tubilidentata, and the aardwolf, a termite eater in the Order Carnivora. All of these animals show a reduction or loss of teeth, probably because the food is small enough to be swallowed whole. The echidnas, giant anteater, tamandua, silky anteater, and pangolins are all toothless. The others have simple, weak, or peglike teeth, generally reduced in number as well. These mammals have powerful forelegs and strong claws for ripping open sun-baked nests. They have long, pointed snouts, narrow heads, and long, tubular, sticky tongues with which they lap up their prey. The South American anteaters have a gizzardlike portion to their stomachs. Armadillos, although not especially myrmecophagous, are primarily invertebrate eaters and show some similar adaptations.

Fish-eating mammals are sometimes highly specialized. Porpoises and dolphins that feed on fish have numerous conical teeth used for holding their prey, which is swallowed whole. The teeth of many seals and sea lions, although not conical, serve a similar purpose. Unlike many of the piscivores, however, fish-eating bats chew their food.

Conical teeth apparently serve equally well for catching squid, as the sperm whales, which feed on this prey, have conical teeth in their lower jaws, but nonfunctional teeth in the upper. Crushing molars are typical of some of the mammals that feed on other mollusks. The walrus rakes up clams and mussels with its long tusks and bristly whiskers and crushes the shells with its strong molar teeth. Other mollusk eaters cope with the hard shells of the prey differently: the sea otter brings its food to the surface and cracks it open with a stone.

One of the more abundant foods in the sea, well used by

mammals, is the euphausiid shrimp. These planktonic forms occur in huge numbers and are the major source of food of most of the whalebone whales. The special adaptation for feeding on this krill is a straining mechanism. The baleen whales lack teeth, and instead have, hanging from the upper jaw's plates of baleen, a skin derivative similar in composition to fingernails. The whale engulfs a mouthful of ocean water containing krill. When it closes its mouth and raises its tongue, the water is forced out past the overlapping, fringe-edged plates of baleen, but the foodstuff remains in the mouth and is swallowed. On a smaller scale, the same process is repeated by the crabeater seal of Antarctic waters, which also feeds on krill. Instead of baleen it has elaborate, scallop-edged molar teeth that serve as strainers.

Blood-eating vampire bats alight on or near their host and walk lightly to the site where they will feed. Vampires, lacking a tail, with a very small membrane on the insides of the legs, and with a long thumb, are able to walk quickly and lightly, unlike most other bats. They have a reduced number of teeth, the existing molars and premolars being very tiny, as well, but the incisors are large, curved, and sharp-edged. With the incisors the vampire makes a shallow cut and then begins to drink the blood as it flows. The blood is drawn up the underside of the tongue, which is curved into an inverted U shape and flows into the mouth over a notched lower lip that, with the tongue, forms a tube. The lips do not touch the host. The stomach of the vampire is very simple, and the digestive tract is relatively short, sufficing for a food that is already reduced by the host's digestion.

Mammals feeding on carrion generally show the same sort of adaptations as carnivores that eat freshly killed prey. However, carrion eaters usually have a good sense of smell and some, like hyenas, have very powerful jaw muscles and teeth, an adaptation for crushing bones.

Omnivores generally do not have extreme adaptations for one kind of food or another. Skunks and racoons, for example, retain the long canines and snipping incisors of their more highly carnivorous relatives but have molars that are fairly flat and can crush and grind, as well as molars and premolars that can shear meat. Most bears have the carnivorous type of incisors and canines but essentially vegetarian-type molars, and they do not have a set of shearing carnassials in their jaws.

FOOD STORAGE

The food supply of many mammals is not constant throughout the year, especially in the temperate and Arctic zones. Fruits, berries, seeds, and nuts may be available only at certain times. Deciduous trees may be bare during the winter. Grass may be green only during the winter. A carnivore's prey may be in hibernation or estivation or may have migrated to a warmer climate or to an area where food is more abundant. For these reasons, many mammals store food.

Storage of live food by carnivores is not common. However, European moles paralyze earthworms by biting them near the anterior end and store them in underground chambers. Short-tailed shrews have been reported to store snails and even beetles alive, although bitten, in their underground chambers. Perhaps the shrew's venom is used to produce paralysis in some animals so that they can be stored alive.

More usually, dead animals are stored. Some weasels, mountain lions, bears, foxes, and wolves are known to put away surplus food, at least temporarily. While the puma will cover a kill with brush and return to it nightly until it is consumed, red foxes are known to make caches of their surplus, but only to return to them in time of need. The same has been reported for wolves and some weasels.

Plant products are the food items most commonly stored, especially by rodents. Gray squirrels are well known for their summer and fall activities of burying nuts. These food supplies are used during the winter and are located then by the squirrel's keen sense of smell. Chickarees are noted for their middens, great storage heaps of pine cones that are buried in damp earth until needed. Most mice cache seeds, nuts, and berries in underground chambers, and some store tubers and beans. They may even put aside cut stems, roots, and shoots, but these probably do not last too long. Beavers put sticks and even small trees under water where they will be available during the winter beneath the ice.

Special treatment of food for storage is not very common, but is practiced by a few animals. Kangaroo rats place fresh seed pods in dirt-covered pits about an inch deep around the opening of the burrow. These are left in the hot, dry soil for about two months, after which they have lost enough moisture to be taken to the underground burrow chambers and kept without danger of mold. Red squirrels hang mushrooms in crotches of trees to dry and cure and later store them in the ground or in a stump.

The pika of the mountains of western North America is well known as a hay maker. In late summer these lagomorphs cut plants from the mountain meadows and lay them out on rocks, stumps, and other exposed places to dry. When the hay is cured, the pikas carry it to shelter under a rock or in a crevice, where it will be available during the long winter.

In addition to oxygen, water, and food, mammals also require certain minerals. A lack of them may result in deficiencies and abnormalities, changed metabolism, and eventual death. For example, insufficient calcium and phosphorus, which are needed for bone formation, will lead to malformations. Other minerals and salts are needed in minute quantities, but are nevertheless essential to the well-being of the mammal. Vitamins, too, are essential to normal health, growth, and reproduction, but relatively little is known of the vitamin requirements of wild mammals.

6 ❦ Defense and Protection

No ANIMAL is wholly safe from attack or aggression from members of its own or of other species. Herbivores are preyed upon by carnivores, small carnivores are bullied and eaten by large ones, and large carnivores are in danger from others of their kind and from the defensive activities of their prey. All mammals are in potential danger from attacks by man, and man may be in danger from attacks by other animals as well as by his own species.

In order to survive, mammals have developed varied defenses. It is said of human wars that a new weapon or defense is completely successful only the first time that it is used, for thereafter a defense or an attack is devised to counteract the novelty. Over the course of evolution, the same appears to be true of mammalian defense and attack mechanisms.

Carnivores are equipped for killing, and the same adaptations that they use to kill their prey generally serve in their own defense. Specialized defensive mechanisms, particularly among the larger-sized ones, are not common. Herbivores, on the other hand, are adapted to feeding on nonmoving, relatively unresisting foods, and have evolved special adaptations to defend against predators.

FLIGHT

One of the most common defensive measures is running away from the danger. The physiological differences between actual combat and fleeing are slight, and the entire reaction has been termed "fight or flight." Fear or anger stimulates the adrenal glands, and from the medulla of these glands are produced secretions called epinephrin and nor-epinephrin, which, when released into the blood stream, produce immediate effects including: contraction of smooth muscle fibers in some blood vessels, causing a rise in blood pressure; relaxation of some blood vessels leading to skeletal (voluntary) muscles; heart stimulation; relaxation of the smooth muscles of the bronchioles of the lungs; stoppage of muscular contractions of the intestine. The total effect is to prepare the animal for fighting or fleeing. Blood is drawn from the viscera and fed to the muscles and brain, the heart rate increases, the lungs can take in more oxygen, and it will be carried faster to the tissues. Even the contraction of skin muscles that causes hair to stand erect is produced by adrenal action, so the animal not only looks larger, but may also cause an attacker to misgauge the location of a vital part when striking.

Speed by a fleeing animal is, of course, necessary, but endurance may not be. It may only be necessary to make a short dash to reach the burrow or other shelter. Similarly, an erratic flight path may be far more beneficial than a straight one. Kangaroos are able to bound off almost at right angles while going at high speed, and the effect of this usually is to put a

DEFENSE AND PROTECTION

closely pursuing predator, which could not alter its course so quickly, a few paces farther behind.

Most mammals that flee to escape enemies have a distance threshold that triggers their flight. Predators in the distance may be ignored, or watched periodically. If they come closer, the prey may stop its other activities and watch, and when the predator decreases the distance to the flight threshold, the prey turns and flees. The distance up to which a predator is allowed to approach varies with the species of prey, and among individuals as well.

A strange result of this flight threshold has worked to the detriment of the American pronghorns. These animals are the fastest runners in North America and live in open country. They have good eyesight and undoubtedly distinguish between different kinds of predators in determining how close they can permit one to come. The necessity for identification of the approaching threat has led to curiosity on the part of the pronghorns, and they are attracted to strange objects, sometimes coming to within fifty feet of a fluttering handkerchief. While their great speed makes them relatively safe from almost any other predator, man has utilized their curiosity to attract them to within shooting range.

Carnivores feeding on speedy animals are generally fast runners themselves, and some carnivores, too, have little endurance and rely on a short, quick dash to capture their food, while others count on endurance to wear down their prey. Most of the members of the cat family rely on a short, quick dash or a sudden pounce from concealment to make their kills. On the other hand, members of the dog family, such as Cape hunting dogs or wolves, usually hunt in packs, and rely on perseverance to run down animals. They often hunt in relays so that when one group of them tires, others take up the chase.

IMMOBILITY

Although flight from enemies seems to be more common, many mammals rely on remaining motionless as a means of escape. This may involve other factors such as coloration and odor, but it can also function without regard to them. Many predators are "sight hunters," meaning that they must see their prey. Most of these mammals are attracted by movement, and a motionless animal may be passed closely by a predator without being observed. Fawns remain still and quiet when left by their mother and are generally unnoticed by predators. (They also are virtually odorless.) Rabbits "freeze" and may be ignored; if approached beyond the flight threshold they will then run. An extreme of passivity is shown by the North American opossum, which feigns death. This action is not merely a lack of motion; the animal becomes limp, the tongue lolls, and the opossum can be moved about without reacting. This lack of activity when moved or touched apparently reduces killing attempts on the part of the predator, and while some individuals "playing possum" are surely killed, many others must escape because of this reaction. Whether "playing possum" is an imitation of death or an actual case of shock is still not clearly determined. Opossums also seem to be distasteful to some predators.

SIZE

A few mammals, because of sheer size, are relatively free from predation. Adult elephants and rhinoceroses are rarely

DEFENSE AND PROTECTION

attacked by any animal other than man. In addition to their thick skin, the tusks or horns can be formidable weapons, and the elephant's trunk may be used in many defensive maneuvers. Each of these animals is so large that any attacker is in danger of being crushed if the animal should step or kneel on it.

Yet the large baleen whales are not safe from predation by killer whales and, except by flight, do not defend themselves. In fact, it seems that they are almost paralyzed by fright when attacked by killer whales, for they allow the killers to butt open their mouths and devour their tongues. It is not known what predators, if any, there are of killer whales; the young of all these large animals, otherwise protected by bulk, are vulnerable.

SKIN

The thick skins of elephants and rhinoceroses are defensive mechanisms, and some mammals have evolved even more elaborate external protection. Armadillos are covered with a bony carapace that affords excellent protection. Some have a large, solid shield that covers the back of the animal like the shell of a turtle, while others have segmented sections between solid front and rear shells, which enable the animal to curl up. One (*Tolypeutes*) can roll up completely, its tail and head shields fitting together to complete a bony sphere. Pangolins, too, are armored. Their heads, tails, and backs are covered with overlapping horny scales. When frightened, these animals roll up with the head tucked in the belly and expose their sharp-edged scales to the attacker.

QUILLS

Porcupines, hedgehogs, and some other rodents, monotremes, and insectivores derive protection from a body covering of hairs that are modified to stiff, sharp quills. Those of the North American porcupine have the tip barbed for a portion of its length, so when the sharp point penetrates the skin of an attacker, the quill detaches from the follicle. The shape of the barbs is such that the quills are difficult to remove and, by the normal contractions of the muscles in which they are imbedded, they work deeper and deeper into the flesh. The quills of hedgehogs, echidnas, and some of the other rodents do not detach from the follicle so easily and are not barbed. Despite what may seem to be a very formidable defense, quilled animals are preyed upon. In America, porcupines seem to make up a good portion of the diet of fishers, which even eat the quills, although how they manage to accomplish this remains to be discovered. None of the quilled mammals, of course, is able to "shoot" or "throw" its quills.

COLORATION

Coloration is an important protective mechanism. Most mammals, with the exception of man, most other primates, and a few others, are color-blind and see things in various shades of gray. Color-blind mammals do, of course, distinguish tones and intensities of color, and some of the predators of mammals, especially birds, do have color vision.

Mammals, as a rule, match or blend in with their natural

DEFENSE AND PROTECTION

background, usually the substrate, or soil, so they can conceal themselves more easily. This matching of the substrate is sufficiently universal so that the generalization can be made that mammals from warm, humid areas are generally dark, and those from hot, dry areas are generally pale. This coloration is essentially to aid concealment rather than a response to temperature and humidity, although these, too, may play a part. In any event, when the factors of concealing coloration and response to climate come into conflict, it seems that concealing coloration prevails.

Perhaps the outstanding example of matching substrate occurs in several populations of pocket mice. Essentially inhabitants of arid lands, most pocket mice are some shade of tan or brown. However, in the Tularosa Basin of New Mexico there are extensive isolated black lava flows. Nearby are white gypsum sands, and elsewhere there are pale brown desert sands. The pocket mice on each of these backgrounds match them extremely well. An almost black pocket mouse lives on the lava, an almost white one on the gypsum sands, and a sand-colored one on the deserts. The climate in each of these places is hot and dry.

Experiments have shown how important it is that a mouse match its background. Using deer mice of two shades of color as the prey, and barn owls as the predators, an artificial background was made with soil that matched the coloration of one of the kinds of mice. Vertical sticks and horizontal cross pieces gave some protection from the birds. Over a range of very low light intensities from dim light to almost total darkness (0.24 foot-candles to 0.000,000,080 foot-candles), it was found that almost twice as many of the "conspicuous" mice were caught by the owls than those that blended with their background (124 to 68), indicating that concealing coloration is of survival value to a mouse, even at night.[1]

Many predators also show concealing coloration, and it is

obvious that the harder it is for a predator to be seen by its prey, the more successful the predator is likely to be in obtaining food.

A number of mammals have concealing coloration that, out of the normal environment, may seem to be quite striking. The young of many kinds of deer are spotted with white on a tan or brown background. This dappled effect not only breaks up the outline of the animal, but simulates the patches of light shining onto a forest floor. Other forest animals are spotted, including pacas, baby tapirs, leopards, and jaguars. Striping, as in tigers, and blotched reticulations, as in giraffes, are different patterns of disruption that enable both predator and prey to blend in with their background.

In Arctic regions some mammals, such as polar bears and Peary caribou, are white throughout the year while Arctic foxes, Arctic hares, varying hares, some lemmings, and some weasels molt in the fall from their brownish or grayish summer pelage to a white winter coat. While there are individual differences and species differences, the environmental stimulus for the coat change is not wholly temperature, but seems instead to be correlated, at least in part, with the decreasing length of day in the fall and the increasing length of day in the spring.

One of the most peculiar arrangements for concealing coloration among all mammals is found in the sloths. These tree dwellers of the American tropics have green algae growing on their hairs, which gives them a mossy tinge that probably aids in concealment. The hairs of one genus are modified with longitudinal grooves in which the plant grows, while the other genus has irregularities that run across the width of the hair where the alga may find purchase.

A few mammals have a coloration that apparently is designed to make them noticeable. Mammals using a noxious odor for protection—three kinds of skunks in North America,

DEFENSE AND PROTECTION

several types of weasels in Africa, and the Asian teledu—are all marked with contrasting black-and-white patterns. Although these patterns may serve functions other than recognition, being for some, such as the spotted skunks, a disruptive, concealing pattern, they serve also as a conspicuous marking, a warning to predators. The black-and-white markings of striped skunks form the greatest contrast that can be viewed by a color-blind mammal.

NOXIOUS ODORS

Odors that are unpleasant to other animals are not limited to the skunks, as they are found in most members of the weasel and mongoose families. Most probably, these odors are used for communication by the relatively less specialized musk bearers, but on occasion some protection may be afforded by an ability to spray a foul-smelling substance. The odors of the skin glands of some shrews seem to deter cats from eating them, but not from killing them. The meat of aardvarks is said to taste of formic acid (from ants), which might serve to prevent man, at least, from killing them, except in starvation situations. So far as is known, no mammal has poisonous flesh, although at certain times of the year polar-bear liver is toxic to humans because of the extremely high concentrations of vitamin A.

OTHER SPECIALIZED DEFENSES

Noise

The noises that mammals make can also serve as protection. A shrill scream made by a rabbit as it is grasped may startle an attacker into dropping it, and conversely the same scream may attract other predators from several miles away—the principle that has led to the use of "predator calls" by hunters. Warning sounds are made not only by the voice, but in different ways. Skunks stamp their front feet as a warning, and kangaroo rats "drum" with their feet. In general, however, vocal calls or screams are probably used more for defense against, or communication to, other members of the same species, as in the threatening growls of members of the dog families or the "saber-rattling" thrashing of antlers in brush by moose.

Teeth

The prime function of teeth is to secure food and to reduce it to small enough pieces for swallowing and digestion. Teeth, however, do function in defense and protection and in a few cases have become highly specialized for these purposes. The tusks of elephants, for example, serve some food-getting function but can be used for defense and protection. Their development as weapons, however, has probably taken place as a part of the *intra*species sociology of these animals, for the tusks are often small or absent in females and therefore presumably have little survival value. Their use by the males, however, may be (or may have been) largely in connection with male rivalry. The spearlike tusk of the male narwhal would seem to be a superb weapon, except that these animals are not known to use it in any fashion for offense or defense. Again, the absence of sizable tusks in the females suggests that this is merely a secondary sexual character, perhaps now a relict of a day when males used them as weapons against one another.

DEFENSE AND PROTECTION

Hoofs and Claws

The hoofs of ungulates can be excellent defensive weapons. The hind-foot kick of a zebra or an ass can stun or even kill a predator, and a moose can use its front hoofs to fight off and kill wolves. Hoofs are used offensively only in battles with others of the same species but are a potential defensive weapon for all those animals that possess them.

Claws are used as offensive weapons mainly by members of the cat family and are used to hold prey or for other purposes by other carnivores. As defensive weapons, they are widely employed. The raking swipe of the powerful clawed paws of a bear is deadly. The kick of the clawed hind feet of a kangaroo can disembowel any Australian predator, and the toothless giant anteater can rake open a dog with a single swipe of its long, sharp claws. Smaller mammals sometimes will roll over onto their backs when fighting in order to bring all four clawed feet into the battle. Even such a supposedly timid animal as a hare can rip open a man's arm with a powerful backward kick of its clawed hind feet.

Horns and Antlers

Antlers, being shed annually, cannot serve for defense and protection throughout the year. Furthermore, except for the reindeer or caribou, there are no species in which the females normally have antlers. Antlers are essentially a secondary sexual appendage of the males and are used mainly in fighting males of the same species during the mating season. When present, of course, they can be and are used as weapons against predators, but in many species the antlers are dropped in winter, in temperate zones at least, at which time the deer would seem to have most need of them because of the lack of shelter from foliage, because snow hampers their escape and because this is the period of least available food for most carnivores.

Horns are permanent structures and are not shed annually. Often they are borne by both the males and females of a species. They are used in battles for mates by males and also for defense by both males and females. The larger mammals with horns use them well, and water buffaloes are reputed to have killed attacking tigers with their horns and hoofs. Nevertheless, the horns of some antelopes have taken bizarre shapes that would seem to preclude their use as defensive weapons of reasonable importance. However, much remains to be learned of the life histories of most of the antelopes, and it may well be that the shapes of the horns have a particular defensive value that pertains to the specialized attacking techniques of some of their predators. It may also be that the horns serve a function of physiological importance in the hot savannas—they may provide additional surface area for cooling heat-loss.

Poison

Poisons are not common among the mammals. The male platypus is known to have a hollow spur on each of its hind legs and a poison sac beneath the skin of the thigh. It can, through these equivalents of a poisonous snake's fangs, inject a venom into an attacker. There are few data concerning the effects of this venom, but it is potent enough to incapacitate a man for some weeks. The presence of this poison apparatus in the males only, however, suggests that it serves, or once served, a secondary sexual function. As pointed out in the preceding chapter, venomous saliva is known to be present in the mouth of the short-tailed shrew and is present in some other shrews. The venom causes convulsions and death in mice and is reported to have caused a man, bitten on the arm, discomfort for a week.[2]

Warning Display

Warning and threat display is shown by many kinds of mammals. The snarling, tooth-revealing grimace of cats and dogs is one example. Hoofed animals may paw the ground prior to a charge, and the raising of the tail of a skunk is a warning of the malodorous spray to come. Spotted skunks may make an even more striking warning display by balancing on their front feet and advancing on the opponent. Llamas spit a foul-smelling saliva when irritated, and some monkeys may, in excitement, break and throw twigs or even their own feces at an intruder. Gorillas thump their chests with cupped palms as a threat. Most of these warning displays are used both against members of the same species and those of other species.

CLIMATIC PROTECTION

The role of shelter as protection from climatic stresses has already been discussed in Chapter 4. It is not the sole means available to mammals, of course. As protection from cold, mammals have hair as a body covering. The effect of this insulation may be very great, and a shorn guinea pig, for example, may lose one-third more heat than when it has its hair. When shorn, a domestic cat at 68°F. loses 16 per cent more heat than it produces, indicating that it cannot long survive without the insulating protection of its fur. The dense fur of Arctic mammals is excellent protection. Arctic foxes, with a fur thickness of about 2 inches, have been shown to exist at −22°F. temperatures without raising their metabolism above the basal level, and it is believed that they begin to lose heat sufficiently to cause their metabolism to increase only at temperatures below −40°. By contrast, a squirrel with about a third of an inch of insulating hair must begin to produce more heat at a temperature of 35° or 40°. Tropical mammals, such as sloths and douroucouli monkeys, must begin to produce more metabolic heat when the temperature drops to 75°. In water much of the insulating effect of hair may be lost, and the mammals that spend most of their lives in water generally have thick layers of fat beneath the skin that function as insulation.

When wet to the skin, the normally nonaquatic mammals lose heat rapidly as a rule. The smaller ones, when soaked, will soon die from heat loss in this condition at otherwise comfortable temperatures. Mammals that are semiaquatic, even in the tropics, generally have a dense coat of short hair that retains air in the spaces and thus provides some protection from water reaching the skin. In addition, the production of oils from the skin glands is sufficient to keep the hair from becoming waterlogged, but not so much that it becomes matted.

The cetaceans and sirenians, which spend their entire lives in water, are virtually hairless. For some time it was thought that the dense-furred aquatic mammals in contrast had to come out on land periodically to dry their hair. While this may possibly be so, a large number of seals and sea lions spend so many months in the water that it would seem that it is not a requisite for these animals. The same is true of the sea otters, which do not often venture onto land. Exceptions to the dense-furred aquatic animals are found among a few tropical and semitropical species. Hippopotamuses have no hairy protection and come out on land to feed at night and to bask by day, and the capybara has a coarse, scant outer coat and no soft, dense underfur.

Hair serves as protection not only from cold, but also to some extent from the hot rays of the sun. The insulating effect, while probably necessary, evidently requires some modification, for during their molts, mammals in areas where the climates are markedly different in winter and summer generally have a lighter, scantier pelage during the warmer part of the year. One of the most interesting adaptations to protect an animal from the sun's rays is found in the diurnal African arvicanthine rodents. Beneath the skin of the skull there is a layer of tissue (periosteum) that is pigmented with black, and it is believed that this is an adaptation to protect the brains of these creatures from the direct rays of the equatorial sun. Related nocturnal species lack this pigmented tissue.[3]

HIBERNATION AND ESTIVATION

Although hibernation and estivation are generally thought of as mechanisms enabling an animal to avoid a season of excessive heat or cold, it seems that these dormant states are as much as, or more, related to a time of little or no food as to temperature. There are actually various levels of dormancy, the most extreme being called true hibernation. It involves a marked drop in body temperature, reduced metabolism, and a condition of torpidity, in which the animal has lost sensibility or the power of motion partially or completely.[4]

In temperate climates a few kinds of mammals "den up" during the colder days. Skunks and raccoons, as examples, are not active for days during extremes of cold, remaining in the den, maintaining their normal high body temperature by metabolizing fat. This does not fit the definition of hibernation and is merely a period of inactivity induced by extremely cold weather.

Temperate-zone bears may also be inactive during the winter and do show a tendency toward hibernation in that there is a slight drop in body temperature, metabolism, and sensitivity. American black bears have a decline of about 13°F. below normal in body temperature; their oxygen consumption drops to about 60 to 50 per cent of normal, and general metabolism is reduced by about one-third or even half. The amount of torpidity of the denned bear is variable, but, as some unfortunate persons have learned, these animals can quickly become active when disturbed. Further, the level of metabolism is sufficiently high so that the gestation of the young takes place during this time in the females. Males are often active during warmer days in the winter, and polar bear males and unbred females do not den up at all during the winter. The type of behavior exhibited by temperate-zone bears is generally called winter dormant.

Of the true, or deep hibernators, only a few have been studied in great detail. Not all of them are identical in their winter torpidity. Not only are the temperatures that induce hibernation or awakening different, but also the entire course of the hibernation varies with species. In general, however, the physiological activities in these periods are similar in all.

While most of the hibernators put on great layers of fat prior to hibernating, a few require even additional insurance. Hamsters, for example, must have a stored supply of food, and they awaken during the winter and feed. Animals that have not achieved sufficient fat are active later in the year than those that have, presumably trying to put on sufficient fat before be-

DEFENSE AND PROTECTION

coming torpid. Animals without enough fat to last through the winter may awaken earlier in the spring and find themselves without a food supply and no fat reserves.

The rate of metabolism of a hibernator may be as little as 1/75 its active rate. The body temperature may drop from a normal of about 95° to 36°F., and the heart rate from more than 100 beats per minute to only 4 or 5. The rate of breathing shows a similar reduction down to one or even fewer breaths per minute. The intestine is inactive, and brain activity is scarcely measurable. The number of white blood cells drops to about one-third the amount during the active state, and the number of blood platelets may double, although disease resistance is increased and clotting time is lengthened.

During the period of hibernation, sleep is not continuous. There are periods of awakening and return to hibernation. These times of sensibility may vary from a few days to a few weeks, but as the end of the hibernation period approaches, the awakenings become more frequent. There is also a critical temperature above which hibernators become active. This varies with the species, from a low of about 50° for the European hamster to a high of about 80° for some European bats. Extreme cold will also awaken hibernators. During hibernation the animal's body temperature varies with the temperature of the hibernation chamber, generally remaining a degree or two above it. However, as freezing is approached, the metabolism of the hibernator increases and it may awaken before its tissues freeze. In this way, it may have an opportunity to move to a deeper chamber or to remain warm for a short period when the ambient temperature, that of the medium around the animal, is lethally cold. The long-term survival value of such a mechanism, however, does not seem apparent, for the animal would not have sufficient fat reserves to live through a long period of cold weather.

The amount of weight lost during hibernation varies with the species, the amount of fat put on during the summer and fall, and the length of the hibernation period. For woodchucks, about one-third the total body weight may be lost, which includes almost 100 per cent of the stored fat reserves. Most hibernators emerge with some fat energy still in reserve, for they often do not awaken during a season of maximum food supply, and there may be a period of time when their food intake does not equal their energy expenditure. During this time there is maximal use of the last of the food reserves in the body. As the season of abundant food approaches, these animals expend their energy on reproduction, and it is only after this activity that the major fat depositions for the coming hibernation are made.

Despite the many studies of hibernating mammals, much remains to be learned. For instance, hibernators show a remarkable resistance to various diseases and drugs, but the mechanism of such resistance is as yet unknown.

Akin to hibernation, and perhaps identical to it, is estivation. This summer torpidity has been far less studied than has hibernation. The inducement for estivation seems to be the disappearance of green vegetation early in the summer in, for example, some parts of western United States. The Washington ground squirrel (*Spermophilus washingtoni*) emerges from hibernation in late January or early February and breeds, the young being born in February or March. By the end of June the adults have gone into estivation, and, without emerging from their burrows, go into hibernation, remaining in a state of torpor for seven or eight months. The young may remain above ground a few weeks later in the summer than the adults, presumably to put on enough fat for the long dormant period.

The various methods of defense and protection mentioned in this chapter are only a part of the entire picture. The home

DEFENSE AND PROTECTION

provides shelter from enemies and climate, skin adaptations provide protection and aid in food-getting, and migration, hibernation, and estivation are mechanisms for survival during periods of inclement weather and short food supply. The social interactions of mammals are a means, also, of survival not only of the individual, but of the species.

7 ❧ Social Structure and Populations

SOCIAL STRUCTURE

No MAMMAL'S LIFE is wholly devoid of contact with other members of its species. The first and simplest social relationship is the one that exists between the newborn mammal and its mother. This relationship may last only until the offspring is weaned (as among gray seals when the baby is left by the mother after two weeks). In any event, no mammal lacks a conscious awareness of another of its own kind.

Those that are commonly said to be "solitary" do spend most of their lives out of direct contact with others of their species. The simplest type of adult social interaction occurs with these solitary species. It consists of little more than a male seeking out a female during the breeding season, reception by the female, and departure by the male. This seems to be the case with many rodents and some carnivores, such as the Canada lynx. Courtship, if any, may occupy very little time, and thus, for the major part of the year, the males, at least, lead a lone existence.

The females, even of solitary species, have more social contact because of the presence of their young. This social relationship may be short or long: the black rhinoceros female may have her young with her for as long as two years. More complex than the simple one-to-one relationship of the rhinoceros is that of the mother and several offspring. Here there is not only interaction between the mother and each newborn, but also between the mother and the babies as a group as well as social relationships of the offspring to each other. This is the type of situation that exists among bears, skunks, raccoons, and many other kinds of mammals.

A further stage in complexity of social organization is developed when the male remains with the female and assists in the raising of the young. The family of male, female, and offspring may persist only until the young are partly raised, as among river otters, or until the young disperse, as is characteristic of red foxes. With the latter the same male and female may mate again the following year, which is just a stage short of the permanent monogamous situation that may be characteristic of some beavers and some wolves.

Social Hierarchy

Within any group of mammals some degree of social hierarchy probably exists, but the subject has not yet been thoroughly studied. One of the major problems in trying to understand the social structure of wild mammals is the need for the observer to be able to identify each individual in the group. Most research, therefore, has been done with laboratory animals or with wild mammals under somewhat controlled conditions. The principles determined from the few studies that have been made probably apply to some extent to most mammalian aggregations.

The social relationships that exist in a group fall into several categories: male-male, male-female, male-young, female-female, female-young, and young-young. One common type of behavior is a dominance-subordination organization in which the relationship of individuals forms a structural social hierarchy. This type of relationship is often called a peck order, because it was first described in chickens and was manifest to a large extent by which individual pecked which other chickens.

A simple, "straight-line" peck order has been observed in a captive aoudad herd consisting of four males, four females, and four young. Dominance was shown by an actual or threatened butt with the head and was determined by giving two individuals equal opportunity to obtain a piece of food. The dominant individual always got the food. Among the males, male *A* was dominant over males *B*, *C*, and *D*; male *B* was dominant over males *C* and *D*; and male *C* was dominant over male *D*. The same linear arrangement of dominance was demonstrated by the relationships of the four females to each other, and the young also showed a linear dominance hierarchy. All males were dominant over all females, so male *D*, subordinate to all other males, dominated female *A* and all the other females. Similarly, all of the females were dominant over all of the young.[1]

Not all dominance relationships are simple linear arrangements, and a triangular organization may exist in which animal *A* is dominant over *B* but subordinate to *C*, with *C* being subordinate to *B*. As the number of animals in a group increases, the social complexity becomes even greater. The formula for determining the number of interactions possible is: number of animals (N) times the number of animals minus one ($N-1$), divided by two, or $\frac{(N)(N-1)}{2}$. Thus, in a group of four animals, the number of possible relationships is 6, but in a herd of 20, the number of interactions is 190, and in a group of 100 the total is 4,950. The larger the group of

SOCIAL STRUCTURE AND POPULATIONS

animals, the less likelihood there is of a simple linear dominance hierarchy.

Furthermore, dominance in one facet of life (food-getting, for example), may not mean dominance in some other aspect. In a pride of lions most of the kills are made by the females. The males, however, are dominant in feeding on the kill. The young can engage in activities with adults (attacking the parent's tail) that would not be tolerated from another adult. No matter what the normal relationship, almost all females become dominant when protecting their young from either males or other females.

One of the classical studies of the social organization of wild animals is that of Professor C. R. Carpenter, an animal psychologist now at Pennsylvania State University, who studied howler monkeys on Barro Colorado Island, a research station in the Panama Canal Zone. The size of the groups ranged from 4 to 35 (therefore 6 to 595 possible interrelationships) with an average of about 18 in each group (153 possible interrelationships). The typical unit was made up of 3 males, 7 females, and 6 infants and juveniles.[2]

Within each group of monkeys there were subgroups, such as a female with her young, defensive groupings of two males and sometimes a female, and others. The entire structure of a howler clan shows a nondominant type of organization. The male-male relationships are devoid of fights. The individual males do not seem to show dominance, one leads at one time, another at some other occasion. The males do seem to stick together, and they do lead the troop with females and young bringing up the rear. When another clan of howlers comes near, or an individual from a different group approaches, the males join in howling at (and in this way driving away) the intruders. Similarly there is no fighting over mates.

The relationship of the males to the females also shows no dominance. As a female comes into heat she approaches a male who stays with her until he is sexually satiated, whereupon the female moves on to another male. The relationship of the males to young is generally one of indifference. However, should a baby fall from a tree, the males commence howling and continue until it is rescued. A male will even retrieve a fallen baby if the mother is not able to.

Female-to-female relationships are similar to those of male to male. There is no fighting, and the females tend to remain together and follow the lead of the males when the group moves. The social interaction between the female and her young is very strong. For the first year the mother is constantly with it, caring for it, sheltering it from rain or cool weather, feeding it, and carrying it. The young monkeys play together, and some slight form of dominance seems to be evident in their biting and wrestling, the only form of fighting howler monkeys engage in.

Much more dominance is shown by gorillas, in which an adult silver-backed male is the leader and is dominant over all others in the group. A linear hierarchy exists between the silver-backed males, all of them being dominant over black-backed (younger) males and the females. The relationship between black-backed males and females varies, dominance working in either direction. All adults are dominant over all juveniles and infants away from their mothers. Females with small infants are dominant over females with larger infants, and both are dominant over females without infants. Among the juveniles and infants, size seems to be the main criterion for dominance—the larger being in power over the smaller.

The social structure of huge aggregations of animals, such as the caribou of Canada or some of the antelopes of East Africa, appears not to be a superorganization, but to consist of small herds, which do, however, generally have a social structure. It may be that there is also some hierarchy between the various smaller groups making up the larger herd, but study of such relationships is difficult and has not yet been thoroughly undertaken.

Certainly social interactions vary with time, age of individuals, temperature and other factors. Group composition changes seasonally; some bats form "nursery colonies" of pregnant females and later of mothers and their young, while the males may be solitary at that same season. Bachelor herds are formed by younger male vicuñas and a similar segregation is noted among Alaskan fur seals during the season of birth and breeding. Striped skunks do not seem to be highly social animals, yet as many as twenty of both sexes have been pulled from a winter den where they had congregated. Some of the marsupials, kangaroos for example, seem to gather in unorganized aggregations with little social interaction, while most of the others seem to show little social behavior beyond defensive behavior or limited maternal activities. The social organization of whales, porpoises, and dolphins appears to be high in some species, but its internal mechanisms are still not understood. Except for studies on howler monkeys and spider monkeys, practically nothing is known of the details of the social behavior of wild South American primates. There are no detailed studies of this aspect of the lives of bats, the second largest group of mammals; and for the largest order, the rodents, the few studies available are of some of the larger, generally diurnal, or noticeably social animals such as prairie dogs.

Communication

In order to have social behavior within a species of mammals there must be communication. This does not mean, as is unfortunately construed by the popular press, that there must be language by which specific information can be transmitted from one individual to another. Communication among mammals may take numerous forms and different senses are used. Sight, sound, odor, and touch are employed, and there may well be others of which we do not yet have knowledge.

There is communication even among the solitary mammals. If they are territorial, and most of them seem to be, there must be some manner of delimiting the territory. This may take the form of an odor marking, usually performed by urinating or defecating at "signposts," at which another individual can learn of the presence of the territory holder without actually encountering him. Specialized odor glands, as in the members of the weasel family, can serve to make these markings, which seem to be as effective as "no trespassing" signs are for humans. A female in heat communicates to males, by means of a distinctive odor, that she is in such a state. The leg, hoof, and facial glands of many kinds of ungulates emit odors used for communication. In some cases they are used to mark territories, but what other information they may communicate is not known.

Sight communication is of great importance, especially to diurnal mammals. The American pronghorn communicates the presence of danger to other members of the herd by raising the white hairs of its rump, heliographing the information. A similar adaptation is found in the African springbuck, which has a fold of skin containing white hairs on its back. The raised, white-tipped tail of a skunk is a warning, mainly to animals of other species, but also to its own. More subtle visual communications are relayed by specific behavior. The dominance hierarchy of wolves, for example, once established, is maintained by facial expressions, positions of the ears, showing of teeth, crouching, or position of the tail. Each of these actions conveys as specific a bit of information as the stylized bowing, or kneeling, or kissing the hand before a

SOCIAL STRUCTURE AND POPULATIONS

king does among humans. Man, a visual animal himself, perhaps understands visual communications better than he does most of the other types used by mammals. The pawing of the ground by an angry bull transmits a threat to a human far more effectively than a written notice.

Sound-hearing communication is also widely used among mammals. Earlier in this chapter the effectiveness of howling by the South American howler monkeys was mentioned. Snarling, growling, and barking by members of the dog family are all facets of sound-hearing communication, and one that man also understands to some extent. The whimpering of its young gives information to a mother, or at least attracts her attention to her offspring. The roar of a stag red deer communicates a challenge to other males and is used only during the rutting season. A hind, by a staccato bark, warns the herd to be alert when something out of the ordinary attracts her attention. The roaring of bull fur seals is a major portion of the territory defense during the breeding season.

The bats and cetaceans, and probably other mammals, including shrews, use a sound mechanism for orientation. The animal emits sounds and, from analysis of the returning echoes, can avoid obstacles or find food. In bats, the level of these sounds is generally above the hearing ability of man, but in the cetaceans, some of the sounds are audible—so much so that the belugas or white whales are known to their hunters as "sea canaries." Dolphins have received a great deal of publicity in recent years because of their highly developed echo-location mechanism. It now seems likely that these animals communicate with one another by their sounds, but this mode of communication is not unique. Other forms are known among mammals, including the kissing of prairie dogs and the grooming of monkeys, and there are undoubtedly types that remain to be discovered.

Value of Social Behavior

In order to evaluate the importance of social behavior to mammals, a philosophical point must be considered: what is the purpose of any animal? This subject cannot be dealt with fully in a book of this length or scope. However, from the evolutionary viewpoint of biology, it would seem that the adaptations of a species are to provide it with mechanisms to *survive* and to *reproduce*. Social structure and communication are methods that aid mammals in these objectives.

A social organization can provide for a division of labor so that the activities of one individual can assist in the survival of the group. A pronghorn antelope, seeing an approaching wolf, signals, by its flashing rump patch, to the other members of its herd the proximity of potential danger. In saving itself in this situation, it also saves other members of the species. The social organization of a troop of baboons is arranged so that the larger and stronger males are placed on the periphery of the troop. An enemy thus immediately faces those members of the group best suited for defense. The females and young have the opportunity to escape while receiving the protection of the males. With baboons, as in any polygamous society, the majority of males are expendable.

The group social organization may function toward survival and reproduction in a somewhat different fashion. A peck order is a means of using a symbol to avoid expenditure of energy. Two males, for example, having fought once to establish which is dominant, no longer need to fight each time a competitive situation arises. Since such competition could theoretically come up every time food is available, the danger of injury or death to one or both competitors would exist daily to the over-all detriment of the herd, to say nothing of the individuals involved. Within the social structure, a ritual battle can be undertaken with a snarl, shaking of the head, or pawing of the ground. This involves much less loss of energy, which may be critical at a time of little food, and reduces the danger of the loss of a member of the herd through death or injury.

In battling for mates, the males of many species may engage in somewhat more than ritual battles, but even in these physical combats there is a degree of psychological advantage to the owner of a territory or one that has previously exerted dominance over the other, so the battle is rarely to the point of death or severe injury.

Social structure and communication, then, are a means of fostering survival of a species through allowing the survival of individuals and groups by decreased energy expenditure and a division of labor.

POPULATIONS

There are several definitions of *population* used not only in biology but in other fields as well. There are both abstract and concrete connotations to these definitions. A useful definition is: The organisms, collectively, inhabiting an area or region. A more specific biological definition is: "a group of living individuals set in a frame that is limited and defined in respect of both time and space." [3] Thus, the meadow mice occupying a specific acre on a research station on July 27, 1968, are a population. By the same token, the meadow mice of the State of New York are also a population.

Being composed of living things, a population has characteristics of its own, some of which are typical of life itself. It has a definite structure and composition which are constant

BIOLOGY OF MAMMALS

at any specific time, but which vary with age. A population of mice has a social structure and a composition of males, females, adults, and young existing at any given moment. However, a month later the social structure and composition will have changed because of deaths, births, and aging of individuals.

The effect of a population on the environment may be greater than the sum of its individuals. Just as the shade and climatic effect of a forest of tall trees is greater than the sum of the effect of individual trees, so the effect of a population of mammals may be greater than the sum of the individuals. For example, one deer in an isolated forest probably could not manage to find and consume every tree seedling, so the forest would continue to reproduce itself. But a population of deer could conceivably eat every seedling and cause a cessation of tree reproduction in that forest.

A population grows in much the same fashion as an animal. Starting rather slowly, the growth accelerates and then levels off, forming a sigmoid curve when graphed against time. If we assume a hypothetical mammal that is monogamous and produces two young once a year, which are able to reproduce in a little less than a year, a growth curve of this population can be constructed. Starting at year zero are the single adult male and female. The following year there are 4 in the population, the original pair and their offspring, one of each sex. The year afterward there would be 4 adults and 4 young, for a population of 8. A year afterward the population will have reached 16, and the following year 32, and it will continue to double each year. Theoretically, this population would go on increasing indefinitely. The potential for reproduction of all mammals is enormous. Meadow mice usually have from 5 to 8 young in a litter. The females can breed at an age of 25 days, and the gestation period is about 20 days. If a litter of 6 is produced every month, and if the litters are evenly divided between males and females, the reproductive potential of a single pair of meadow mice is more than 6,500 in a year. When one considers that 15 meadow mice per acre is considered to be a low population, it is soon evident that the theoretical reproductive potential of a single low population acre of these animals is about 45,500 mice in a year, or, put in weight, a total of more than 2½ tons of mice. Even elephants, which have a gestation period of nearly 2 years and which are reproductively mature at an age of 12 could theoretically produce a herd of 20 from a single adult pair in 24 years.

It is quite obvious that the theoretical reproductive potential of any animal is rarely reached or, if it is reached, not long sustained. Most mammals maintain a population level that is in equilibrium with their environment, sometimes increasing, sometimes decreasing, but generally fluctuating within limits that are neither the theoretical maximum nor minimum.

The size of a population is governed by a relationship between the birth and mortality rates. If the birth rate increases and the mortality rate remains constant, there will be an increase in population size. If there is a constant birth rate and a decrease in mortality rate, there will also be an increase in population. Should the mortality rate increase without an increase in birth rate, the size of the population will decrease. Numerous factors govern alterations in natality or mortality, and these govern population fluctuations.

Factors Favoring Increase in Population

Influences favoring population growth can be of either genetic or ecologic origin, or both. Genetic factors would include a hereditary change of birth rate tending toward larger litters, shorter gestation period with more frequent litters, or an increase in reproductive season over a given span of time. All of these would produce more young in a given period of time. For instance, an animal that ordinarily has only a single litter a year might have a mutation that would enable it to produce a second litter a year, thus increasing the birth rate.

The ecologic factors, of course, affect the genetic factors as well as having an over-all effect on population size. A genetic change to produce a second litter might seem advantageous, but it would actually be deleterious for the population as a whole if the young were born in a time of year when the climate was severe and the young, less able to survive because of their age, might pre-empt den sites or food critical for the population as a whole. Thus, a second litter a year might cause a reduction in the population because of the increased mortality resulting from the additional, unsupportable members born into the group.

Usually ecologic effects on populations are considered under two headings: those that result from the population size itself and those that are independent of the population. Density-dependent factors favoring an increase in population would include such items as available space. If the population is increasing and the space for nests or dens is completely occupied, then the increase must stop. But if space is available, then the conditions created by the animals themselves may make an optimal environment in which to increase their numbers. For populations, however, density-dependent factors as a fostering influence are possibly not as important for increase in numbers as they are for causing the decline of a population.

Density-independent factors include climate, water, total space available, and food production. They are extremely important as influents for the increase of population. A prolonged warm spell in the autumn may give a late litter a better chance to survive the coming winter by making less energy demands. A heavy snow early in the winter may protect hibernating and subterranean animals from later freezing because of its effect of insulating the ground from extreme cold. Exceptionally good seed crops in the fall may assure survival of a greater than normal number of animals through the winter. Similarly, a good spring plant growth may encourage earlier and more prolific breeding of mammals because of the greater available nutrition.

Factors Favoring Decrease in Population

The same factors, genetic and ecologic, can work toward a decrease in reproductive rate or an increase in mortality to cause a decrease in population. Moreover, the ecologic factors are either density independent or density dependent. In general, there is better knowledge of the factors that cause a population to diminish in size than there is of those that result in an increase. The reasons for this difference in information is that large populations are more noticeable, the cause of death of individuals often can be determined, and there is an economic importance attached to knowing how to control animals that reach high populations. (There is also an economic importance to the knowledge of the factors that cause increase in populations, but because such knowledge involves foresight and anticipation, it is rarely pursued.)

The genetic factors that cause a reduction in population include such aspects as hereditary decrease in litter size, low viability of young, sterility, and increased prevalence of lethal genes in a population. Density-independent factors such as climate can cause population reduction by such means as drought, flood, food-crop failure, excessive heat or cold, and lack of snow cover during the coldest parts of the year.

The density-dependent factors that favor decrease in population size are generally interrelated and often quite subtle.

When a population is very large, many factors that normally are inconsequential take on a major importance. When a population is low, there is more space available for each animal, while with a high population there is crowding. Some individuals must take less favorable homesites—places where they may more likely be subjected to flooding, drought, or exposure to predators. Similarly there is less feeding area for each animal as the population increases, with the result that individuals may suffer from malnutrition or starvation. Even if they are not starving, they may have less fat as a food reserve, and an early winter, prolonged cold snap, or other climatic factor could provide just that extra drain on the energy of the animals to cause their death.

As a population increases, its predators generally increase as well. In some cases the carnivores seem attracted by the food and move in from far and wide. In the case of mice, if the population is large enough, coyotes, foxes, weasels, hawks, owls, snakes, and other predacious animals move in quickly. The carnivore population begins to increase either because of larger numbers of its young being born or because of better survival of young resulting from the more abundant food. Predation then becomes a major factor in causing the reduction of a population.

The crowded conditions resulting from a large population provide another mortality factor by increasing the transmissibility of disease and parasites. The greater the number of contacts among the members of the population, the greater are the chances of transmitting diseases that, if not fatal, can weaken animals so that they have less of a chance to escape predators or to obtain food or mates in competition with others of their kind. Parasites, too, are more widely spread by a large population. Generally they do not kill the host, but they do serve to weaken it, or at least require it to obtain more food to make up for the nutrition taken by the parasites; and

SOCIAL STRUCTURE AND POPULATIONS

in crowded conditions there is more competition for food, which the parasite-ridden animals are possibly less capable of obtaining than those that do not have parasites.

One of the most interesting of the density-dependent factors is a direct result of the crowding that takes place in high populations. As a result of the competition for food, shelter, mates, and the stimulus of repeated contact with others of the same species, physiological changes take place in the animal. The constant stress under which it is living causes continual activity of the cortex of the adrenal gland (see Chapter 6, page 86). The effects of this overstimulation of the adrenal cortex may cause a reduction in the number of young produced. Although the number conceived may be as great as at any time, a greater number of the embryos are resorbed by the uterus, so reproduction in the population can virtually cease, causing the population to decline. Another facet of this adrenal-cortex syndrome is to cause depletion of the cortex to produce what has come to be called "shock disease." In extremely high populations it has been noted that an animal startled by an investigator may run a few paces and fall over, dying. Autopsies have indicated that the adrenal cortex as well as sugar reserves were exhausted; and when called upon to prepare the animal for the excitement stimulated by the investigator, the internal chemistry of the body could produce no epinephrin or energy, and the animal went into a state of shock, coma, and died.

It is precisely this type of stress that a social structure helps to alleviate. The peck-order social hierarchy tends to reduce the total amount of stress by letting each animal know its place. Those higher up in the hierarchy, because of the lessened stress, tend to produce more young. The animal at the bottom of the peck order may not be able to reproduce at all. That much of this stress is of psychological origin may be evident from experiments in which mice were permitted to breed in a cage until it was quite crowded, when it was noted that breeding had ceased and the population remained at a high level. A peck order was present in the population as determined by competition at the food and water troughs, but production of young had ceased at every level. However, when a tranquilizer was introduced into the drinking water, reproduction started again. In other words, the stress of crowding no longer bothered the animals, and the factors that induced reproductive inhibition no longer were present.[4]

In summary, the benefits of social organization to the species as a whole are fairly evident. If nothing more, it provides a greater degree of assurance that one pair of individuals, at least, will get enough food, protection, and minimum stress to continue to produce young and thus continue the species. Even in such things as resistance to disease it has been found that the dominant individuals in a population have greater resistance than do those lower in the social hierarchy. Essentially, social organization is a means of energy conservation.

Population Cycles

It has long been noted that some animals go through periodic increases and declines of population with a certain regularity. The causes of these cycles are still being sought, but much of the information presented earlier in this chapter was developed in the course of studying mammalian population cycles. One of the best bodies of information on cycles is the fur records of the Hudson's Bay Company of Canada on snowshoe hares and lynxes. For more than a hundred years these two animals have been reaching population peaks and declines at approximately ten-year intervals. Lynxes feed almost entirely on snowshoe hares, at least when they are available, but the peaks of population do not coincide. The highest snowshoe hare year is followed by a great decline in hare numbers the following year (there are about ten times as many hares at the peak population as there are the following year). The lynxes reach their peak population in the year when the hares are at their lowest, and the following year there is a great drop in the number of lynxes, which find themselves starving for lack of hares. Then the hare numbers begin to build up the following year and the subsequent years, reaching a peak some six to nine years after the previous peak.[5]

Cycles are normal for many of the smaller rodents, especially lemmings and meadow mice, which reach peaks every three to four years. At times, with the great reproductive potential of mammals, there are mass outbreaks of mice, such as the one of European house mice that occurred in California in the winter of 1941–42. It was estimated that there were 17 mice per square yard, 82,000 per acre; damage to crops during such irruptions is devastating. Disease, shortage of food and shelter, and death from control operations and predation caused a quick decline in the number of mice by the following year.

A variety of ideas about the causes of cycles have been presented. The hare-lynx cycle has been attributed to sun spots (which occur in a cycle that does not always coincide with the peaks and valleys of the hare-lynx cycle). It seems probable that there is no one specific "cause" of such cyclical phenomena in mammals but that the multiple factors that encourage reproduction and reduce mortality when the population is at its lowest are followed by years of increasing numbers until the peak is reached, when the accumulation of factors that favor a decline precipitate a "crash" in the population. The intervals between maximum populations would be attributed to the length of time that it takes the various kinds of mammals to reach their greatest numbers, at which time the factors favoring decline come into noticeable play.

Mammalian populations are rarely in a natural balance, but rather are fluctuating from year to year. They may range from a low of merely a single breeding pair of animals, to a high that taxes the carrying capacity of the environment, and the animals may, indeed, eat up all the available food so that most of the population then dies of starvation. These fluctuations cannot be viewed as a self-contained entity, but must be considered in relation to the entire environment, for the effects of one species are inextricably enmeshed with the lives of the plants, other animals, and the structure of the physical environment in which it lives.

8 ❧ Mating, Reproduction, Gestation

From the evolutionary viewpoint of biology, the function of an animal is to produce others of its kind, as was stated in Chapter 2, where birth was called an evolutionary end, and reproduction is the means of achieving it. Mammalian reproduction is unique, although most of the features, taken singly, can be found among other classes of vertebrates: internal fertilization in reptiles, placentation in sharks, and parental care in birds. But taken together these characteristics have given the mammals a method of assuring the production and survival of their offspring.

REPRODUCTIVE SYSTEM

There are three basic types of mammalian reproduction, but each is accomplished with essentially the same anatomical structure. The primary differences are in the amount of development that takes place within the body of the mother before the young are born.

BIOLOGY OF MAMMALS

The male reproductive system consists of two testes in which the spermatozoa are produced. A group of small tubules, the epididymis, through which the sperm pass from the testes, lead to a larger duct for each testis, the vas deferens. These vasa deferentia join in a common duct, the urethra, which also carries the urinary excretions from the bladder to the outside. Various glands (seminal vesicles, prostate, bulbourethral or Cowper's) add secretions to the spermatozoa; the resultant solution, called semen, aids in the transfer of sperm to the female. The urethra passes through the penis, which is the organ by which the sperm are introduced into the body of the female.

There are, of course, modifications of this basic anatomical system in the various groups of mammals. The testes, for example, are housed permanently in a skin-covered sac, the scrotum, located below the anus in most of the hoofed mammals and primates. The scrotum, for those mammals that have one, is generally behind the penis but in the marsupials is anterior to it. In rodents, the testes are generally within the body cavity, but descend into the scrotum during the breeding season. Bats and some of the insectivores lack a scrotum, but the testes come to lie nearer the surface of the body during the breeding season. Whales, seals, elephants, and rhinoceroses have testes that are permanently within the body cavity.

The epididymis not only transports spermatozoa from the testis to the vas deferens, it also serves as a temporary storage area, and some bats may hold the sperm over the winter in a special pouch. Some maturation of the spermatozoa also takes place in the epididymis. The various accessory glands differ in location in the various kinds of mammals, and all of them are not present in all species. The penis, which is the intromittent organ of mammals, varies greatly in size, shape, and structure, according to the kind of animal involved. In the platypus it seems to be nothing more than a simple groove on the floor of the cloaca, a common chamber for the products of the reproductive, excretory, and digestive systems. Many rodents, bats, carnivores, and some other mammals have a bone, the os penis or baculum, in the penis, and the varied shape of this structure has been widely used as a means of classifying and aging the animals that possess it. The penis of marsupials is forked.

The female reproductive system basically consists of two ovaries which produce the eggs (ova) and a pair of ducts, the oviducts, which are the tubes down which the ova pass into another pair of tubes, the uteri, sometimes fused into a single structure. The uterus opens to the outside by way of another duct, the vagina. The way in which these organs function, and their arrangement, however, differs in three major ways, as mentioned earlier.

The egg-laying mammals, the platypus and the echidnas, do not utilize the uterus, which is small and little more than a junction of the oviducts. The fertilized egg develops as it passes down the oviduct, has a leathery shell added, and is deposited into a urogenital sinus, a common chamber for the products of the excretory and reproductive systems. From there it passes to the outside through the cloaca. Most of the development of the young, then, takes place outside the body of the mother, either in the platypus' nest or the echidna's pouch.

The marsupials show an advance over the condition found in the egg-laying monotremes by having the reproductive system separated from the digestive system, although there is still a common chamber for the products of reproduction and excretion. The fertilized egg passes down the oviducts and the uteri into a sac formed in marsupials from paired vaginas, which is called the vaginal sinus. It is here that the little internal development of marsupials takes place. Although the normal passageway for the young to be born in this case would seem to be through one or both vaginas, this does not occur. Instead, the vaginal sinus dissolves or ruptures and forms a median tube that permits the fetus to pass down into the urogenital sinus and then to the outside. Among the marsupials, there is variation in this system, with some maintaining the median vaginal tube for the rest of their lives after the first birth, and others healing the wound and rupturing the vaginal sinus anew for each birth.

The third type of female reproductive system is that which occurs in all the other mammals, broadly called the placentals. The placental reproductive system has only a single vagina and a greater separation of the urinary and reproductive tracts. Placentals vary considerably in the kinds of uteri, and they are usually categorized into four major types. In the duplex uterus, each horn is separate and each has a separate opening, the cervix, to the vagina; this type is found in the rodents and rabbits. Most of the carnivores and some of the ruminants have a bipartite uterus, in which the lower portions of the two uteri are fused, and have but a single cervix opening into the vagina. The bicornuate uterus that occurs in many of the hoofed animals has a much greater fusion of the lower portion of the two uteri, being about two-thirds of the organ. Functionally the difference between the bipartite and bicornuate uteri is that in the former the implantation of the embryo or embryos occurs in the separate portions of the uteri, while in the latter implantation takes place in the fused, lower section. The fourth type, the simplex uterus, is found in the higher primates, and here both uteri are completely fused to form a single chamber.

Although female reproductive organs are paired, or at least bilaterally symmetrical, they are not necessarily functionally so. In the platypus only the left ovary is functional, and eggs have been found only on the left side of these animals. In some bats only the right ovary is functional, and embryos are found implanted only in the right horn of the uterus. In other bats both ovaries are functional, but the embryos have been found in the right horn in 70 per cent of the gravid females.

SECONDARY SEXUAL CHARACTERS

Besides the internal anatomical differences between the sexes, the males and females of many kinds of mammals show differences that are not primarily related to reproduction. These are called secondary sexual characters, and they are generally controlled by the various hormones released by the anterior portion of the pituitary gland and glandular cells of the gonads, the testes, and the ovaries.

The presence of antlers in males of the deer family is a male secondary sexual character. These growths appear because of certain hormonal influences, and castration of white-tailed deer while the antlers are present results in their being shed within a month. The next annual growth of the antler takes place at the normal time, but the shape of the antlers is abnormal, and the "velvet," the soft tissue that covers and nourishes the antlers as they are growing, is not shed. Female deer (except in the caribou) are antlered only abnormally, generally a result of pseudohermaphroditism, a condition in which an animal that is genetically a female has some male organs and a confused anatomy and endocrinology.

The brightly colored rump patches of males of some species of Old World monkeys, and the bright skin of the face of mandrills are secondary sexual characters. The poison spurs on the hind feet of platypuses, the long tusks of narwhals, the inflatable proboscis of elephant seals, and the mane of African lions are other examples of male secondary sexual characters. Many males are larger than the females, but some rabbits and whales, for example, have females whose average size is larger than that of the males.

SEXUAL CYCLES

In the wild, breeding periods of individual mammals are generally restricted to limited times that are governed both by the receptivity of the female and the sexual preparedness of the male. The period of receptivity by the female is called the estrus, heat, or rut. Species that have only one heat period each year are said to be monestrus, as exemplified by some temperate-region bats, deer, bears, and most members of the dog family. Mammals that come into heat several times a year are polyestrus. If the animal comes into heat several times during a restricted portion of the year, it is said to be seasonally polyestrus. Cottontail rabbits (*Sylvilagus floridanus*) are seasonally polyestrus, breeding from mid-January or February until July or August, and have several litters during this period. Many domestic mammals are polyestrus, coming into heat throughout the year.

The regulation of the period of heat is under the control of several hormones that maintain a delicate balance within the animal. During anestrus, the sexually quiescent period before the breeding season, the female is generally not interested in mating, and the reproductive organs show no special state relating to breeding. As the breeding season approaches, the anterior portion of the pituitary gland produces a hormone that stimulates the development of the ova. The ova (or ovum in mammals that produce only one at a time) lie on the surface of the ovary in a fluid-filled chamber, known as a follicle. The stimulating hormone from the anterior pituitary is called follicular stimulating hormone, generally abbreviated to

F.S.H. A second pituitary hormone, luteinizing hormone (L.H.), furthers the development of the follicular ovum and probably influences the rupture of the follicle at the time of ovulation. In addition, the F.S.H. stimulus causes the ovary itself to produce a group of female hormones, estrogens. Their action is added to that of F.S.H. and L.H. to mature the ova. As the level of estrogens in the system increases and ovulation nears, the secretion of F.S.H. begins to diminish.

With the release of the egg from the ovary, the follicle cells form a new and temporary gland, the corpus luteum (literally, "yellow body"). The secretions of this gland, progestins, are active in pregnancy and birth. One of the main functions of the progestins is to stimulate the walls of the uterus to prepare for the implantation of the embryo, should the ovum be fertilized. At the same time, progestins also act with estrogens to suppress further the ovary-stimulating production of F.S.H. The estrogens also function in readying the uterus for the forthcoming pregnancy.

If the egg is not fertilized, the corpus luteum may disappear, and, with the inhibiting influence of the progestins and estrogens removed, the anterior pituitary can begin the cycle again, or the animal can go into the quiescent anestrus once more until the next breeding season. In some mammals, however, the corpus luteum may persist even if fertilization has not occurred, and the female goes through a pseudopregnancy. In this condition, the animal may go through all of the physiological and behavioral reactions of pregnancy (even to the extent of building a nest and producing milk) in the same time sequence that occurs in a normal pregnancy. Domestic dogs often may have a pseudopregnancy of varying duration, and unfertilized domestic rabbits can undergo one that lasts about half the normal gestation period.

If the ovum is fertilized, it is later implanted in or on the wall of the uterus where the protective membranes (amnion, chorion, yolk sac, and allantois) and the placenta (through which nutrients and oxygen are passed from the bloodstream of the mother to that of the embryo) are formed under the influence of progestins. Some of the membranes, the placenta, and the uterus themselves also secrete hormones, which function in a fashion similar to the luteinizing hormone. As the time of birth nears, another hormone from the corpus luteum, relaxin, is produced. This serves to relax the ligaments joining the two sides of the pelvis so that the opening through which the fetus must pass can spread slightly and facilitate the passage.

At the time of birth, the inhibiting effect of estrogens on the anterior pituitary ceases because there is a sharp drop in estrogen production, and the hormone in the system disappears rapidly. Another pituitary hormone, galactogen, or prolactin, stimulates milk secretion in the mammary glands, which have previously enlarged under the influence of this same hormone.

The endocrine interactions that take place in the female reproductive cycle are actually far more complicated than outlined here. Several organs or glands may produce the same hormone at the same or at different times. The removal of a gland may terminate a pregnancy in one species and seemingly not affect the process in another. The same type of hormone may act as a stimulant to a gland at one stage of the cycle and an inhibitor later.

A reproductive cycle in a male mammal follows much the same sequence as that of the female, so far as appropriate. The follicular stimulating and luteinizing hormones from the anterior pituitary stimulate testis tissues to produce male hormones, androgens, principally testosterone. This hormone not only affects sperm formation, but has a great influence on the secondary sexual characters of the males, and its secretions are responsible in part at least for antler formation in deer, swelling of the neck during the rut in many hoofed animals, and the general state of sexual excitement that occurs at the time of breeding. Eventually, the androgens inhibit the pituitary's secretion of F.S.H. and L.H., and a decline takes place in the sexual organs of the male until the cycle starts again before the next breeding season.

As outlined here, both the male and female reproductive cycles start with secretions from the pituitary. But what stimulates the pituitary? For animals that are polyestrus throughout the year, it is very likely that the pituitary responds to the rise and fall of the amount of various hormones in the blood. In polyestrus mammals, progestin from the corpus luteum inhibits the production of follicular stimulating hormone from the anterior pituitary. If the female is unfertilized, the corpus luteum disappears, and the anterior pituitary is free again to produce F.S.H. The length of time that it takes for the diminution of the corpus luteum to the point of no progestin secretion would seem to be one of the main factors that governs the length of time between heats in the polyestrus animals. This, however, is not the whole story, for some polyestrus animals do not respond in this fashion, and it has been shown that some of the glands involved in the ovarian cycle undergo a refractory period, a time during which they cannot be stimulated in the normal hormonal fashion.

Seasonal breeders, either monestrus or polyestrus, may also have a refractory period between the last cycle of one season and the first of the next breeding, but some seasonal breeders are strongly influenced by changes in the external environment. Many of the species that are seasonally polyestrus in the wild are potentially year-round breeders. For example, the American bison in natural conditions bred during a restricted period, July and August, but in captivity bisons can breed at any time of the year. It is thought that in the wild, the scarcity of food and unfavorable climate during the winter inhibited pituitary stimulation, while in zoos, where the animals are well fed and perhaps sheltered, there is no such inhibition. Another example is the white-footed mouse (*Peromyscus leucopus*) which is seasonally polyestrus (April to October) in the northern part of its range (northeastern and northcentral U.S.), but in the southern part it breeds throughout the year.

In temperate regions most mammals breed during the warmer portions of the year, from late winter to early autumn. In the tropics, breeding seasons, in general, are less restricted, and individuals of many species may be breeding at any time of the year. In some cases, however, breeding seasons seem to be related to the annual cycle of rainy and dry seasons. The South American tapir (*T. terrestris*) is said to breed just before the beginning of the rainy season, and as more is learned about the life histories of tropical mammals, others will undoubtedly show such seasonal correlations. The mechanism by which such climatic factors stimulate or inhibit the production of F.S.H. from the pituitary, however, is not known.

The one environmental factor that has been studied most in relation to sexual cycles is photoperiod—the proportion of light and dark each day, which becomes more pronounced with distance from the equator. It is not surprising, then, to find a number of northern mammals that respond to photoperiod. (This factor is largely uninvestigated for high-latitudes native mammals in the southern hemisphere.) European ferrets normally start to breed in March or April, but, if they are kept under conditions, beginning in October, where the light duration each day is increased gradually instead of following its normal decrease, they come into heat three months earlier than usual. If the animals are kept in total darkness, they do not come into heat at the normal time. Immature ferrets kept on an artificial "short day" take longer to reach puberty than usual, and those that are kept on an artificial "long day" reach puberty earlier than expected.

Similarly, domestic sheep, which tend to breed in the autumn (on a decreasing photoperiod), can be brought into breeding condition in the spring by altering photoperiod. Photoperiod-responsive animals transferred from one hemisphere to another change their breeding cycle to conform with their new environment. Although little testing has been done on the equator, where photoperiod differences do not exist, it is known that some ferrets there maintained the breeding cycle of the area from which they had come.[1]

The mechanism by which light influences the secretions of the pituitary gland is not established to the satisfaction of all investigators. It had originally been assumed that the influence of light on the eye itself was transmitted by the optic nerve to the brain and then to the pituitary. However, nervous connections to the anterior pituitary have not been found, and further, the removal of the eye and exposure of the optic nerve to light produces the same response as when the eye is present, although without the retina of the eye the optic nerve does not transmit impulses stimulated by light. It has been established (with birds) that light can penetrate the skull and reach the anterior pituitary directly. However, mammals tend to have modified to avoid the penetration of light into the body (see comments on p. 99 concerning the arvicanthine diurnal rodents of Africa), and many of the responsive species are nocturnal and are not out in the daylight. Much more research is needed into this aspect of mammalian biology, and it is clearly evident that the photoperiod response of one species cannot automatically be assumed to be the same in other species, or possibly even among members of the same species.

In general, then, the seasonal reproductive cycles of mammals are the results of the combinations of influences of environmental and internal stimuli that may include light, temperature, food supply, and psychological and sociological states, to mention but a few.

COURTSHIP AND COPULATION

Courtship is a preliminary to copulation and varies from very simple and brief activities to complex and lengthy ones. Complete observations of these facets of the lives of mammals are lacking for most species in the wild, and it is well known that sexual behavior in captivity often deviates markedly from the norm. Courtship and mating require that the participants be in the proper physiological and psychological state. A female that is not in heat is ignored by a sexually active male. Receptivity by the female is a necessity for most mammalian mating. Rape (in the sense of a forceful copulation with an unreceptive female) seems notably a human action, or at most one confined to the higher primates.

Courtship activity may be initiated by the male or by the female, depending upon the species. In many small rodents the female runs from the initial attentions (sniffing or licking of her genital region) and is chased by the male, but in some kangaroo rats (*D. ordii*) the female nuzzles the male's genitals. Rubbing of the genitals, posturing, crouching, urinating on objects or the mate are all courtship attractants used by one sex or the other. Circling, prancing, and ritualistic dancing are recorded for many kinds of antelopes and deer. Characteristic mating noises are made by sexually active animals such as bottlenose dolphins.

Mating in some species may take place again almost immediately after the young are born. Alaskan fur seals are bred

MATING, REPRODUCTION, GESTATION

within a few days after they have given birth, and the same is true in many rodents such as rice rats. In these species there is generally a delay in implantation of the zygote, the fertilized egg, discussed later in this chapter.

In a few mammals mating may take place before young are born. European hares (*Lepus timidus*) may show superfetation, in which the female is actually pregnant with two litters at the same time. The same situation seems to occur in some kangaroos, but the complex endocrinological details that permit this have not yet been worked out.

Copulation in most mammals takes place with the male mounted over the back of the female—dorso-ventrally. Belly-to-belly copulation is reported for some species, including hamsters, two-toed sloths, and captive gorillas, chimpanzees, and orangutans. For the gorilla, at least, only dorso-ventral copulation has been observed in the wild. The duration of copulation may be only a few seconds or it may be quite prolonged, and it is highly variable, as when copulation is repeated over a long period of time. A Shaw's jird is reported to have copulated 224 times in two hours, and other rodents may repeat copulation many times in a short period.

In their mating behavior, mammals vary from promiscuity to monogamy. The majority, especially insectivores and rodents, seem to be completely promiscuous. The female mates with the first male to appear when she is receptive, and the relationship between the two lasts no longer than the time required for whatever preliminary courtship may be necessary and for copulation, and then the animals part. Polygamous species have a somewhat longer period of association, but this, too, does not last beyond the breeding season, and the composition of a male's harem the following year may include none of the females of the previous year. Notable among the polygamous species are many kinds of deer and sea lions. Lifelong monogamy seems to be rare among the mammals, although in

135

many species the males and the females may remain together for varying lengths of time after mating. Some kinds of foxes may remain together until the young are raised, but monogamy for more than one or two breeding seasons is rare and has been observed only among wolves, the American beaver, and the lar gibbon.

GESTATION

Gestation is the period of pregnancy, the time from the fertilization of the egg until birth. This varies greatly in mammals, lasting from 8 to 12 days in the marsupial "cat" and 12 days in the American opossum to almost 2 years in elephants. The young of marsupials, as noted in Chapter 2, are born in a comparatively undeveloped state and have relatively short gestation periods.

Among the placental mammals, the shortest gestation period seems to be that of the golden hamster, which can produce young 16 days after mating. Generally, the larger the mammal, the longer the gestation period, but there are so many exceptions to this that another viewpoint can be taken. Animals that produce precocial young usually have longer gestation periods than animals of equivalent size that bear altricial young. A guinea pig has a gestation period of about 64 days and produces a well-developed offspring, with the body well haired, the eyes and ears open, and the youngster ready to move about. A dog also has a gestation period of 64 days, but its pups require a long period of development before they are well haired, open the eyes and ears, and are ready to run around.

There is some relationship, as well, to the size of the litter, those animals with large litters having a shorter gestation period, even within a species. Where only one offspring is produced at a time, there is a slight tendency for the gestation of a male to be longer than that of a female, and the male offspring is usually a little heavier at birth. Within a species there may be variation in the average length of the gestation period; among domestic cattle, for example, Holsteins have an average gestation period of 279 days while brown Swiss cows average 290 days.

It appears that where there is hybridization between two kinds of mammals with different gestation periods, the influence of the mother predominates. Horses have a gestation period of about 335 days (although it varies with the breed), and asses have a gestation period of 365 days. A female horse crossed with an ass has a gestation period of about 345 days, closer to her normal length of pregnancy, but longer than usual. A female ass sired by a horse has a gestation period of 355 days, shorter than normal for the mother, but still closer to that of the ass than to that of the horse.[2]

Some mammals have relatively long gestation periods, but produce altricial young that have relatively short periods of development. These mammals have delayed implantation, a process which delays the development of the fetus until a time closer to the best period for birth. The American marten, one of many members of the weasel family that has delayed implantation, has a gestation period of about 250 days. After mating, usually in July or August, the fertilized egg undergoes some development for a few days and then lies in the uterus, unimplanted until January. At that time this blastocyst is implanted in the uterus and about 50 days of development take place, with birth occurring in March. Thus of the 250-day gestation period, only about 52 days are spent in the actual development of the offspring, and for the remaining period the fertilized egg lies dormant. Delayed implantation occurs in many northern weasels, martens, fishers, bears, some sea lions, and American mink. The process in some, at least, seems to be influenced by photoperiod, for not only is the gestation period for a given species shorter in the lower latitudes (although the period of internal development is the same), but under artificial circumstances, by increasing the amount of light during the fall, the period of delay of the implantation of the blastocyst can be shortened by as much as three months.

Another form of delayed implantation takes place in some of the rodents, should the female be fertilized while she is still suckling a litter. Many rodents of the Family Muridae have a gestation period that is normally 21 days, but which may be extended to 30 or 40 days if mating takes place while the mother is nursing. There is a relationship to the number of young being suckled, and the more sucklings, the longer the delay. It is thought that the lack of estrogens during the period of lactation is the cause of the delay of implantation, for when estrogens are injected into a mated, lactating mother, implantation is prompt.

The bats of the northern temperate zones mate in the fall, but fertilization does not take place until spring. In contrast to delayed implantation, this process is called delayed fertilization. The sperm are stored in the uterus of the female during the winter, while the bats are in hibernation. A secondary mating period may take place in the spring as the bats come out of hibernation.

LITTER SIZE

The number of young that mammals produce varies with the species and ranges from one to twenty or more. A single offspring is characteristic of many of the larger hoofed animals, the higher primates, cetaceans, sloths, elephants, and manatees. The smaller marsupials, insectivores, rodents, lagomorphs, and carnivores generally have three or more young, and a pregnant Malagasy tenrec has been recorded as having 32 embryos. The number of embryos does not necessarily indicate the size of the litter, for frequently embryos are resorbed by the uterus.

Mammals with precocial young usually have smaller litters than those bearing altricial young. In part this is because the longer gestation period and consequent greater size of the fetus does not permit enough space in the mother for a large litter. Precocial young have a better chance of survival because they soon are able to fend for themselves and are not subject to dependence upon the mother for a long period of time. Generally, the number of young is lower in animals with a better survival potential. In mammals that have more than a single offspring at a time, the first litter of a female is generally smaller than the subsequent ones.

Most female mammals breed throughout their lives. Definite information concerning cessation of breeding ability is available only for man, where the menopause (cessation of ovarian function) occurs normally at about 49 years. Should this phenomenon exist in wild species, little has been recorded about it, and probably few females live long enough to reach it. It is known that a captive European brown bear ceased to ovulate at an age of 30 years, and the same is known to have occurred in ranch nutria and laboratory rats at, of course, lesser ages.[3]

Reproduction, then, is the biological means of perpetuation of the species. Mammals are conservative in their reproduction, with each pair producing during their lifetimes more than enough offspring to replace themselves. The longer-lived species usually have smaller litters and less frequent breeding periods than do the short-lived mammals. Protection of the fetus is accomplished (except for monotremes) by a period of internal care followed by a period of parental nourishment and care. With the production of the next generation, the life cycle of a mammal commences once more. There is, of course, far more to the lives of mammals than their individual biologies, for all life is not only related through evolution, but is interrelated through the interactions of animals with one another as well as with the environment. The place of mammals in the economy of nature is the subject of the next chapter.

9 ⌘ The Value of Mammals

THE economic importance of mammals falls into two broad categories—their value in the economy of nature and their direct importance to man. Because of their world-wide distribution, their relatively large size, their abundance, or their activities, mammals are often a dominant feature of the regions they inhabit. As ecological dominants, they may influence the direction of evolution over a relatively large geographic area. While their direct importance to man's economics is better known and evaluated, the subtle, and for the most part unstudied, importance of mammals in the economy of nature is rarely recognized. That the position of mammals in the complexities of the ecological community are only moderately known and that the subtlety of these roles has rarely led man to recognize their importance in no way lessens the actual significance of mammals as ecological dominants. That man chooses to regard mammals only in the light of those whose value may be assessed directly as being good or bad is no credit to man.

INDIRECT VALUE OF MAMMALS

Herbivores

The influence of plant-eating mammals in a region may be very great, although it has never completely been evaluated. The great short-grass prairies of western North America were certainly affected by the millions of bison, prairie dogs, and other herbivores that inhabited them. It is thought by some ecologists that the bison, by their grazing, and the prairie dogs, by their selective feeding, were important factors in maintaining the short-grass prairie. The bison even may have helped to keep back the forest at the eastern extremity of the grassland by their proclivity toward rubbing against, and thus debarking, trees.

Alteration of the environment is most noticeable in the activities of beavers, which not only kill trees by their felling and damming, but create an entirely new habitat that different types of vegetation may invade. Abandoned beaver ponds, silted in, provide fertile soil. Soil turnover is an important attribute of the activities of moles, gophers, prairie dogs, and many other burrowing mammals. By bringing soil up to the surface and through their tunnels, they speed the weathering of the subsoil, and water is conserved because porosity is increased.

Many rodents, lagomorphs, and ungulates damage or kill trees by feeding on bark, but others help to reseed vegetation by burying nuts and seeds. The seeds of many kinds of plants are spread by passing through the digestive tracts of mammals or by being attached to their hair. Pollen-eating and nectar-feeding bats pollinate plants, and the manure of mammals is a fine natural fertilizer. In grasslands, where there are large numbers of ungulates, the soil-enriching effect of mammal droppings is notable.

The brevity of these few examples of the influence of herbivores on the environment must not be construed as being relative to their importance. The subtle influences of these animals are rarely evaluated or recorded, but their importance is great. Wallows on the Great Plains, where bison took mud or dust baths, are still detectable, although a hundred years may have passed since the last bison wallowed there; the further existence of the great saguaro cactus in Arizona is threatened because, despite the thousands of seeds they disseminate, abundant rodents are consuming most of them, and reproduction of these desert plants has almost ceased. If the saguaros disappear, the myriad of animals associated with the plants will also be affected—especially such birds as the gila woodpecker and gilded flicker, which nest in them. The abundance of rodents in this habitat is presumably a result of man's killing of coyotes, foxes, and bobcats, which normally prey on rodents. This is but one of the examples of the interrelationships that may affect the influence of herbivores on the environment. Every species has its own position in the ecological web of the environment.

Carnivores

The importance of carnivores in the undisturbed ecological community cannot be overestimated. One of the classic examples of what happens to the herbivorous mammal population and to the vegetation when the carnivores are removed was demonstrated on the Kaibab Plateau of northern Arizona. There, on some 727,000 acres, 816 pumas and more than 7,000 coyotes were killed between 1907 and 1939, and the timber wolves were exterminated by 1926. This predatory-animal control was undertaken to allow an increase in the mule deer on the plateau, which numbered about 4,000 in 1907. By 1917, when 600 pumas, 3,000 coyotes, and most of the wolves had been eliminated, the deer population had increased to more than 40,000. Damage to the forest was reported in 1917, but the elimination of the predators continued. By 1925 the deer population had reached 100,000, about three times the carrying capacity of the forest, and in the next two winters some 60,000 deer, weakened by lack of food, died. The small trees, shrubs, saplings, and every bit of vegetation that the deer could reach were eaten. The population continued to decline, to about 20,000 in 1931, and the decline persisted, because of lack of food, to 10,000 deer in 1939, despite the fact that during this period the predators were still being eliminated. The population continued downward even after a return to normal was attempted by cessation of the predator-control measures. The ecology of the vegetation of the Kaibab had been changed markedly by this experiment.[1]

A similar case occurred with moose that established themselves on Isle Royale in Lake Superior in 1912. They reached a population of more than 1,500 within a few years, and then, because of overbrowsing, the population crashed to a few hundred in the 1930's. Slowly the population began to build up again as the vegetation recovered, but in 1940 timber wolves reached the island; and in recent years, although there are about 160 moose born each year, the population has held relatively stable at 600 because of the activities of the wolves. Recent studies have shown that most of the winter kills by the wolves are of the sick and weak and that even a large wolf pack finds it difficult, if not impossible, to bring down a strong, healthy, grown moose in the winter.[2]

The activities of coyotes, foxes, weasels, and other carnivores are highly important in controlling the numbers of mice, rats, other rodents, and rabbits. In areas where these smaller carnivores have been eliminated, plagues of rodents or rabbits have often resulted. Studies of carnivores have indicated their importance in removing the sick and weak from populations, which is a major factor in evolution. Even if the prey is healthy, carnivores are a necessity in the natural community to control the numbers of herbivores, and the populations of carnivores are, in turn, regulated by the abundance of the prey species.

As insect-eaters, mammals are of major importance. The insectivorous bats are of inestimable value in this regard, but no detailed study of this important economic facet has ever been made. Ant- and termite-eaters are of great importance in controlling these insects in the tropics. Shrews, abundant though seldom seen, must play a major part in keeping down the numbers of invertebrates.

The effects of baleen whales on marine crustaceans must be of ecological significance, and the effects of other marine mammals such as toothed whales and seals on fish, squid, and octopi is likewise an economic factor little studied from the ecological view.

As the herbivores are of great biological importance in their effects on the vegetation and the soil, so the carnivores are of biological significance to the herbivores. The relationships of these two kinds of feeders have not been well studied from the purely ecological point of view, but there is sufficient knowledge to know that what affects one affects the other—a relationship seldom considered by man in his own attempts to alter the environment.

DIRECT VALUE OF MAMMALS TO MAN

Man's relationships to his fellow mammals have been largely egocentric. His concern with them, even today, is almost entirely from the point of view that mammals are good if they benefit him or bad if they do not. If they do not fit either category they are disregarded. Early man was concerned with mammals only as sources of food and shelter or as potential dangers to himself. The advances of civilization have not altered this concept, except perhaps to broaden it. The acceptance of the viewpoint that other mammals are an integral part of our world has yet to come. In his current activities man protects and fosters those mammals that are of direct value to him and destroys those that threaten him or his wards. The remainder, which make up the bulk of the species, are rarely protected and often destroyed as part of activities to destroy man's "enemies."

Man has domesticated some species of mammals to provide for his use of them more conveniently. This has led to less concern about the wild progenitors of some of these mammals, with their subsequent extinction. Domestication is a separate and distinct aspect of man's utilization of mammals and will therefore be dealt with as a distinct unit.

Wild Mammals

FOOD

With increasing numbers of domestic mammals and with the advances in transportation that have occurred in the past half century, there has been a decrease in the dependency of mankind upon wild mammals as a major source of food. The wild caribou herds of Canada are still essential to the lives of the Indians; catching marine mammals is basic to the Eskimo's survival; and kangaroos and wallabies are important to the existence of the Australian aborigines. Although most hunting in "civilized" countries is now considered a sport, many

THE VALUE OF MAMMALS

hunters utilize the meat of their game for food, especially such mammals as rabbits, squirrels, deer, and wapiti.

Hunting mammals for food is usually on a relatively small-scale, local basis, largely because the areas where it is done are relatively remote and technologically unsophisticated. In the hunting of whales, however, the effort is large-scale and technologically advanced, and the techniques used are such as to assure the extermination of many of the larger kinds of whales if the current quotas are not reduced.

SKINS

At the present time, wild mammals are probably utilized more for their skins than for food. The scanty hair covering of man probably led him to utilize the skins of mammals for protection relatively early. Wild mammal skins have, however, taken on cultural values, such as the association of ermine with royalty, or mink and chinchilla with opulence, and thus the fur trade is subject not to the actual needs of mankind but to social whim. The development of artificial fibers makes it theoretically possible to discontinue the fur trade, but this is not likely so long as the psychological needs of some persons require support in the form of natural furs.

The leading fur bearers in demand and value are members of the weasel family: mink, marten, fisher, otter, and so forth, and aquatic mammals, such as seals and sea lions, muskrat, and nutria. In the United States alone the number of mammals used annually for fur is in excess of twenty-five million, and the value of the manufactured garments approaches a half billion dollars.

In addition to a demand for pelts, leather is made from the skins of some wild mammals, although the bulk of the marketed material comes from domestic stock. Wild boar, peccaries, kangaroos, wallabies, seals, sea lions, and capybaras are hunted for their hides. Novelty items, such as purses and baskets, are even made from the carapaces of armadillos.

OTHER ECONOMIC VALUES

Oil rendered from the fat of mammals was in much greater demand a century ago, before the exploitation of petroleum. Whales are still hunted for their oil, which is used primarily in the manufacture of soap and margarine, in paints, and for softening leather. The famed "sperm oil," a waxy, inedible fat, was used for candle manufacture, but today is used in cosmetics and ointments, as a lubricant for fine machines, and in the manufacture of detergents and shoe polish.

Asian and African civets and American skunks provide musk that is used as a base in the perfume industry. While the skunk musk is a by-product of the fur trade, the Old World civets are often kept in captivity, and their glands are periodically "milked" of scent. Ambergris is a waxy substance that is found in the intestine of some sperm whales. It is believed to be the result of an intestinal disturbance, possibly irritation caused by the beaks of the squids on which these whales feed or by impacted feces. It is used in the perfume industry as a fixative.

Ivory from elephants and walruses is still a valued product, used for piano keys and billiard balls. The development of plastics with equivalent characteristics, however, has greatly reduced the ivory trade. Hippopotamus teeth, narwhal tusks, and the teeth of sperm whales are also sold for their ivory.

In the Orient the horns of rhinoceroses are thought to have an aphrodisiac or medical value, and even small bits of them bring a high price. The same is true of deer antlers in China. The glands of whales are utilized in western medicine for their hormone content, and whale liver supplies large quantities of vitamin A.

Glues and gelatins are prepared from the skeletons of whales and are used in the manufacture of photographic film and candies. Dog food is made from whale meat, seals, sea lions, and kangaroos. Fertilizer is made from the remains of the whaling products, and the guano of bats, dug from the floors of caves, is a rich and valuable fertilizer.

Because of their physiological similarities to man, wild mammals may be infected with diseases that also can infect man. The full extent of this ecological relationship is generally unnoticed until an outbreak of disease occurs in human beings. Some of the diseases (plague and typhus) are transmitted to man through the intermediary of an ectoparasite, while others (rabies and tularemia) may be transmitted directly through bites or contact. Both viral and bacterial diseases may be common to man and other mammals. Certain insect parasites (species of fleas, ticks, bedbugs, and lice) find man as well as other mammals acceptable hosts but generally are only a factor of annoyance to man unless they also carry disease. Some endoparasites (roundworms and flatworms) also find man and other mammals suitable hosts for either intermediary or definitive stages.

The direct danger of wild mammals to man is potentially great, but actually minimal. Unprovoked attacks on man are not numerous, but the larger carnivores, tigers and lions especially, may kill and eat humans. Carnivores also prey on man's livestock, which has led to drastic measures in most parts of the world: the larger and medium-sized carnivores are constantly being hunted and killed today, and the world populations of these species are diminished and, in some cases, in danger of extinction.

The decimation of carnivores generally leads to overpopulations of herbivores in a region, with consequent damage to natural vegetation and man's crops or grazing lands. Some of the hunting activities of man discussed under "sport" are now a necessity to control herbivores from which natural predator restraints have been eliminated.

The greatly increased use of wild mammals for laboratory investigations in medical and psychological studies has, to some extent, been an outgrowth of the air-transport industry. The diseases common to both man and other mammals foster the use of wild mammals for laboratory investigations. Procurement and handling of wild animals for laboratories and for zoos is a small, but widespread, industry, with the potential of overexploitation of wild stocks.

SPORT

Hunting animals as a recreation is a major industry, and men in all parts of the world engage in this activity. Some analyze this sport as part of an inherent "hunting instinct" in man, while others attribute it to an attempt to shore up a weakened sense of virility in a society where the emancipation of women has led to increased competition between the sexes. Whatever the reasons, it is of major economic importance. The basic equipment may be relatively inexpensive, but the subsidiary items expand with the financial means of the hunter. In recent years, in France, because of the decimation of rabbits from the introduced disease myxamatosis, hunting so declined that several manufacturers of shotgun shells were forced to close their factories.

Sport hunting is generally considered in two classes: big game and small game. The latter is usually confined to the smaller mammals, such as rabbits and squirrels, and is done on a local basis. The big-game hunter, on the other hand, usually has to travel far from his home to reach the range of his quarry and consequently incurs considerable expense in travel and lodging, as well as needing more expensive equipment and clothing.

THE VALUE OF MAMMALS

With the decline in numbers of big game in some parts of the world, the esthetic pleasures to be derived from mammals have come to be realized by some people, albeit too small a group. Camera safaris to Africa are now as numerous as the hunting trips, and a growing audience is finally beginning to appreciate the pleasures of seeing animal life in its natural habitat. Unfortunately, this recognition may have come too late for the salvation of the habitat or the mammals, and some may soon be lost to the sight of man forever.

Domestication of Mammals

The great dependency of man on mammals led to domestication of some species to provide a more certain supply of some commodities. There seems to be no wholly adequate definition of what constitutes a domestic animal. It is useful to make a distinction between domesticated and domestic. The former term can be used in the sense of "tamed," and by this definition almost any individual can be domesticated. Domestic animals are a population that, through direct selection by man, has certain inherent morphological, physiological, or behavioral characteristics by which it differs from its ancestral stock.

Most of the common domestic animals came under the direction of man before written history. In some instances their presumed wild progenitors are extinct, and for others the ancestry is not surely established.

The dog was probably one of the first, if not the first, mammal to be domesticated. For primitive hunting men, the dog provided its superior ability to scent game, to warn of dangerous predators, and to track, harass, and kill prey. In addition to their services as hunting aides, dogs have been used for

151

transportation, guarding, herding and protecting livestock, food, skins, and sport. Domestic dogs were probably derived from wolves on a rather local basis, and dogs and wolves still can be crossed with no loss of fertility. Some consider jackals the ancestor of the domestic dog, but there seems to be no evidence for this at present. Just when dogs were first domesticated may never be known, but there is good evidence for their existence as domestic animals at least 9,000 years ago.[3]

Sheep and goats may have been domesticated as early as 9,000 years ago and pigs possibly as early. The domestic goat is believed to have been derived from the Middle Eastern *Capra hircus aegagrus* at least 8,500 years ago. The angora and other long-haired goats may have been derived from the markhor (*Capra falconeri*).

Sheep remains appear in archaeological sites even before goats, and there is evidence from northern Iraq that they were domesticated as early as 10,500 years ago. From just which wild sheep the domestic version was derived is unknown, but the mouflon seems the logical candidate. While goats were probably used primarily for meat and hides and could survive in scrub country, sheep were selected by man for wool and fat. Moreover, as grazers they would not necessarily compete with goats. Sheep today are the most numerous of domestic mammals, numbering nearly one billion.

Cattle were domesticated at least 6,000 years ago. The European cattle are believed to have been derived from the now extinct aurochs and the Asiatic cattle from the banteng. Cattle are the most important of the domestic mammals as a source of meat, milk, and hides today.

The Indian buffalo, or water buffalo, was domesticated at least 4,000 years ago, and the wild stock from which it came and with which it interbreeds freely still exists in places in southeast Asia and India. Its main use is as a draught animal in agriculture.

THE VALUE OF MAMMALS

There is good evidence for domestic pigs 5,800 years ago, but it is believed that domestication took place much earlier in eastern Asia and that the domestic stock was slowly moved westward. An alternative is that the *idea* of domestication of pigs commenced in eastern Asia and traveled westward, local wild stock being domesticated wherever the idea was adopted. A major attribute of pigs as meat producers for the domesticaters must have included their omnivorous habits, for they would not have competed directly with other livestock and would serve as scavengers.

The earliest records of domestic horses are from about 4,000 years ago, and it is thought that domestication took place only about 500 years earlier. Again, although in dispute by students, it seems likely that the idea of domestication of horses spread rapidly and that local wild stock was domesticated wherever the idea met with favor. Riding was a relatively late development and was preceded by cart pulling, and before that, probably, the horse was primarily a source of food. Wild asses possibly were domesticated even before horses; there is evidence for this in southwestern Asia (with the kiang) and North Africa (with the donkey) about 4,500 years ago.

One-humped camels are believed to have been domesticated 4,400 years ago in ancient Babylonia, and their mention in the Old Testament is dated at about 3,800 years ago. The evidence for the domestication of the Bactrian camel is scanty, but it probably occurred at least 2,500 years ago in Persia. The advantages of camels over other domestic stock are largely in their ability to survive in arid regions.

In the New World, two members of the camel family were domesticated by the progenitors of the Incas. Both the llama (as a beast of burden) and the alpaca (as a wool-bearer) are believed to be derived from the guanaco. There is archaeological evidence that they were domesticated at least 3,200 years ago. Another pre-Inca domestication was the guinea pig, the only rodent to be bred for food. When this occurred is not known, but by the time of the Spanish conquest of the Incas, a number of color and hair types had been developed.

The domestic cat is thought to have been derived from one or several local races of wildcat (*Felis sylvestris*). Just when this took place, and even where, is yet to be determined. There is reasonable evidence for domestic cats in Egypt 3,500 years ago. The cat was deified in Egypt, and its domestication probably was in response to its ability to destroy rodents injurious to man's agricultural enterprises. The European ferret, a domesticated form of the polecat, seems to have been used by man for rodent control and for hunting rabbits as early as 2,000 years ago. Another relative latecomer to the domestic scene is the European rabbit, which seems first to have been domesticated in Rome 2,100 years ago.

Man has utilized many other kinds of mammals as either domestic or domesticated animals. Elephants, captured from the wild herds and tamed, are still used as work animals in Asia and in Africa. Cheetahs have been tamed as hunting animals. The position of the reindeer, which has a long history of utilization by man, may qualify it as a domesticated mammal, rather than a domestic one, for there is little evidence of selection by man for specific characters. Locally in Asia domestic yaks and gaurs (called gayals) are utilized, with the wild species still extant.

Man is greatly dependent upon domestic mammals for a great number of products. Dogs, cats, cattle, sheep, goats, and pigs are the most widespread domestic mammals and are generally found in association with civilized man, wherever he lives. Certain other species have been domesticated in relatively recent times to serve specific purposes: laboratory rats (from *Rattus norvegicus*), laboratory mice (from *Mus musculus*), and golden hamsters (from *Mesocricetus auratus*).

THE VALUE OF MAMMALS

MAN AS A MAMMAL

Man is but one of the 4,000-odd species of mammals existing on earth today. The study of him is rarely considered within the province of mammalogy or even zoology and falls to specialists in the field of anthropology. Man's egotism rarely permits him to equate himself with his fellow mammals, although his problems in regard to the topics covered in this book are basically the same as those of other mammals. For the past four hundred years man's mechanical ability and intelligence has enabled him to alter the biotic equilibrium under which he formerly existed, so his survival has become greater than the mortality that formerly kept his population within bounds. This has led to the great human population (3¼ billion persons at this writing) that now inhabits this planet.

Sufficient knowledge is available to inform man of the fate of excessive populations. Should man choose not to regulate his population, he can well expect a combination of physical and biological phenomena to do it for him, with results that surely will alter life on this earth as man now knows it. Man also possesses the means to extinguish himself as a species and may succeed, thus becoming the first creature on earth to become extinct with the knowledge that it was doing so.

THE FUTURE OF MAMMALS

Of the genera of mammals that have existed since this class became distinct from the reptiles, more than two-thirds are extinct. These have died out in the course of the experiments of natural selection, or have faded after spawning new and more successful genera. Just what man's contribution to the extermination of species of mammals was in prehistoric times is not known, but at best it was probably slight and perhaps no more than a hastening of a process under way through the normal course of evolution. In the past two thousand years, however, it is evident that man's activities have caused the extinction of mammals at a highly disproportionate rate. The areas that have suffered most have been islands of isolation where the local mammals could not withstand the pressures of hunting or, more disastrous still, the introduction of man's competitive livestock and his commensal rats and mice. By and large, the majority of extinctions in the past two millennia have resulted not from man's direct activities, but from his *indirect* ones. The feeling for the loss of these creatures, each species unique, the end product of millions of years of experimentation by nature, cannot readily be expressed in words, for it is an emotion. That man, the creature who prides himself on his ability to think, should have needlessly and thoughtlessly wiped out these threads of the fabric of evolution belies his scientific name, *Homo sapiens.*

Mankind stands today at a fork in the biological road. He must decide whether he will live with or without wild mammals. The decision cannot wait, for each hesitant moment brings another species closer to the finality of extinction. Man has the means to exterminate every wild mammal, and he is utilizing that means somewhere on earth every minute of the day. The greatest danger to the existence of wild mammals is no longer the man with the gun, trap, or poison, but explosives, saws, bulldozers, and domestic livestock. Throughout the world man is destroying the habitat of mammals. It does no good to prevent hunting of big game in a forest if the lumberman destroys the home, food, and shelter of animals. It is

THE VALUE OF MAMMALS

thoughtless to set up a wildlife preserve where domestic stock will utilize all of the available drinking water. Is a single domestic sheep, or even a dozen or a hundred or a thousand sheep, worth the knowledge that never again will a human or any other creature see a living Tasmanian wolf?

If man possesses but one iota of the knowledge he professes, then it is surely clear that he need not destroy any more species of mammals. If the human mind can integrate the complexities of physical science that can send a rocket to report information about a planet more than three hundred million miles away, there is no reason why it cannot learn the biological facts necessary to conserve what remains of the earth's mammalian fauna. Man and the other mammals have a common ancestry. Together we have experienced the pressures of evolution. Together and interdependent we have survived. Now that man's dependence on his wild classmates no longer seems necessary, he need not senselessly destroy them.

Man has made a sufficient number of biological mistakes to indicate that, should he choose the fork in the biological road that leads to the extermination of wild mammals, he must remember that he, too, is a mammal and that it may be the end for him as well.

APPENDIX
REFERENCE NOTES
GLOSSARY
SELECTED BIBLIOGRAPHY
INDEX

🐾 🐾 🐾 APPENDIX

Technical Names of Selected Mammals

The following list contains the technical designations of mammals as used by the author. The group names, therefore, may be applicable only to this book and not to mammalogy in general.

AARDVARK	*Orycteropus afer*
AARDWOLF	*Proteles cristatus*
ALPACA	*Lama glama*
ANTEATER, GIANT	*Myrmecophaga tridactyla*
ANTEATER, SILKY	*Cyclopes didactylus*
ANTELOPE	Family Bovidae, several groups
ANTELOPE, PRONGHORN	*Antilocapra americana*
AOUDAD	*Ammotragus lervia*
ARMADILLO	Family Dasypodidae
ARMADILLO, GIANT	*Priodontes giganteus*
ARMADILLO, NINE-BANDED	*Dasypus novemcinctus*
ASS, DOMESTIC	*Equus asinus*

ASS, MONGOLIAN WILD	*Equus hemionus*
AUROCHS	*Bos primigenius*
AYE-AYE	*Daubentonia madagascariensis*
BADGER	Subfamily Melinae
BADGER, AMERICAN	*Taxidea taxus*
BADGER, EUROPEAN	*Meles meles*
BANTENG	*Bos banteng*
BAT	Order Chiroptera
BAT, BIG BROWN	*Eptesicus fuscus*
BAT, GULF FISH-EATING	*Pizonyx vivesi*
BAT, HUMMINGBIRD	Genera *Leptonycteris* and *Choeronycteris*
BAT, VAMPIRE	Family Desmodontidae
BEAR	Family Ursidae
BEAR, BLACK	*Ursus americanus*
BEAR, BROWN	*Ursus arctos*
BEAR, GRIZZLY	*Ursus arctos*
BEAR, POLAR	*Ursus maritimus*
BEAVER, AMERICAN	*Castor canadensis*
BEAVER, EUROPEAN	*Castor fiber*
BINTURONG	*Arctictis binturong*
BISON, AMERICAN	*Bison bison*
BISON, EUROPEAN	*Bison bonasus*
BOBCAT	*Felis rufus*
BEAR, WILD	*Sus scrofa*
BUFFALO, WATER	*Bubalis bubalis*
CAMEL	Genus *Camelus*
CAMEL, BACTRIAN	*Camelus bactrianus*
CAMEL, ONE-HUMPED	*Camelus dromedarius*
CAMEL, RACING	*Camelus dromedarius*
CAPYBARA	*Hydrochoerus hydrochaerus*
CARIBOU	*Rangifer tarandus*

APPENDIX

CAT	Family Felidae
CAT, DOMESTIC	*Felis catus*
CAT, FISHING	*Felis viverrinus*
CAT, HOUSE	*Felis catus*
CAT, MARSUPIAL	*Dasyurus viverrinus*
CATTLE	Tribe Bovini
CATTLE, DOMESTIC	*Bos taurus, Bos indicus*
CHEETAH	*Acinonyx jubatus*
CHICKAREE	*Tamiasciurus douglasii*
CHIMPANZEE	*Chimpansee troglodytes*
CHINCHILLA	*Chinchilla laniger*
CHIPMUNK, EASTERN	*Tamias striatus*
CHIPMUNK, WESTERN	Genus *Eutamias*
CIVET	Family Viverridae, several genera
COATI	Genus *Nasua*
COLUGO	Genus *Cynocephalus*
COTTONTAIL	Genus *Sylvilagus*
COYOTE	*Canis latrans*
DEER	Family Cervidae
DEER, MULE	*Odocoileus hemionus*
DEER, ROE	*Capreolus capreolus*
DEER, WHITE-TAILED	*Odocoileus virginianus*
DESMAN	Subfamily Desmaninae
DOG	Family Canidae
DOG, CAPE HUNTING	*Lycaon pictus*
DOLPHIN	Family Delphinidae
DOLPHIN, AMAZON RIVER	*Inia geoffrensis*
DOLPHIN, BOTTLENOSED	*Tursiops truncatus*
DONKEY	*Equus asinus*
DUGONG	*Dugong dugon*

ECHIDNA	Family Tachyglossidae
ELEPHANT	Family Elephantidae
ELEPHANT, AFRICAN	*Loxodonta africana*
ELEPHANT, ASIATIC	*Elephas maximus*
ERMINE	*Mustela erminea*
FERRET, EUROPEAN	*Mustela putorius*
FISHER	*Martes pennanti*
FOX	Subfamily Caninae, various species
FOX, ARCTIC	*Alopex lagopus*
FOX, GRAY	*Urocyon cinereoargenteus*
FOX, RED	*Vulpes vulpes*
GAUR	*Bos gaurus*
GAYAL	*Bos gaurus*
GAZELLE	Tribe Antilopini, several species
GERBIL	Subfamily Gerbillinae
GERENUK	*Litocranius walleri*
GIBBON	Genus *Hylobates*
GIBBON, LAR	*Hylobates lar*
GIRAFFE	*Giraffa camelopardalis*
GOAT, DOMESTIC	*Capra hircus*
GOPHER	Family Geomyidae
GOPHER, POCKET	Family Geomyidae
GORILLA	*Gorilla gorilla*
GUANACO	*Lama guanicoe*
HAMSTER	Tribe Cricetini
HAMSTER, EUROPEAN	*Cricetus cricetus*
HAMSTER, GOLDEN	*Mesocricetus auratus*
HARE	Genus *Lepus*
HARE, ARCTIC	*Lepus arcticus*
HARE, SNOWSHOE	*Lepus americanus*
HARE, VARYING	*Lepus americanus*

APPENDIX

HEDGEHOG	Family Erinaceidae
HIPPOPOTAMUS	*Hippopotamus amphibius*
HIPPOPOTAMUS, PIGMY	*Choeropsis liberiensis*
HOG, WART	*Phacochoerus aethiopicus*
HORSE	*Equus caballus*
HYENA	Genera *Crocuta* and *Hyaena*
JACKAL	Genus *Canis,* several species
JAGUAR	*Leo onca*
JIRD, SHAW'S	*Meriones shawi*
KANGAROO	Genus *Macropus*
KIANG	*Equus hemionus*
KINKAJOU	*Potos flavus*
KOALA	*Phascolarctos cinereus*
LANGUR	Subfamily Colobinae, several genera
LEMMING, BROWN	*Lemmus trimucronatus*
LEOPARD	*Leo pardus*
LION	*Leo leo*
LION, MOUNTAIN	*Felis concolor*
LLAMA	*Lama glama*
LYNX, CANADA	*Felis lynx*
MANATEE	Genus *Trichechus*
MANDRILL	*Mandrillus sphinx*
MARKHOR	*Capra falconeri*
MARTEN	Genus *Martes,* several species
MARTEN, AMERICAN	*Martes americana*
MIERKAT	*Suricata suricatta*
MINK, AMERICAN	*Mustela vison*
MINK, EUROPEAN	*Mustela lutreola*
MOLE	Family Talpidae
MOLE, EASTERN AMERICAN	*Scalopus aquaticus*

MOLE, EURASIAN	*Talpa europaea*
MOLE, GOLDEN	Family Chrysochloridae
MOLE, MARSUPIAL	Genus *Notoryctes*
MOLE, TOWNSEND'S	*Scapanus townsendii*
MONGOOSE	Tribe Herpestinae
MONKEY, DOUROUCOULI	*Aotus trivirgatus*
MONKEY, HOWLER	Genus *Alouatta*
MOOSE	*Alces alces*
MOUFLON	*Ovis musimon*
MOUSE, DEER	Genus *Peromyscus*
MOUSE, GRASSHOPPER	Genus *Onychomys*
MOUSE, HOUSE	*Mus musculus*
MOUSE, MARSUPIAL HONEY	*Tarsipes spenserae*
MOUSE, MEADOW	Genus *Microtus*
MOUSE, PINE	Genus *Pitymys*
MOUSE, POCKET	Genus *Perognathus*
MOUSE, RED SPRUCE	*Phenacomys longicauda*
MOUSE, WHITE-FOOTED	*Peromyscus leucopus*
MUSKRAT	*Ondatra zibethica*
NARWHAL	*Monodon monoceros*
NUMBAT	Genus *Myrmecobius*
NUTRIA	*Myocastor coypus*
OKAPI	*Okapia johnstoni*
OLINGO	Genus *Bassaricyon*
OPOSSUM, MURINE	Genus *Marmosa*
OPOSSUM, NORTH AMERICAN	*Didelphis marsupialis*
ORANGUTAN	*Pongo pygmaeus*
OTTER	Subfamily Lutrinae
OTTER, GIANT RIVER	*Pteronura brasiliensis*

APPENDIX

OTTER, RIVER	Genus *Lutra,* several species
OTTER, SEA	*Enhydra lutris*
PACA	*Cuniculus paca*
PANGOLIN	Genus *Manis*
PECCARY	Genus *Tayassu*
PHALANGER	Family Phalangeridae
PIG, DOMESTIC	*Sus scrofa*
PIG, GUINEA	*Cavia porcellus*
PIG, WILD	*Sus scrofa*
PIKA	Genus *Ochotona*
PLATYPUS	*Ornithorhynchus anatinus*
POLECAT	*Mustela putorius*
PORCUPINE	Family Erethizontidae
PORCUPINE, NORTH AMERICAN	*Erethizon dorsatum*
PORPOISE	Family Phocoenidae
POSSUM, AUSTRALIAN	Subfamily Phalangerinae, several genera
POSSUM, STRIPED	Genera *Dactylopsila* and *Dactylonax*
PRAIRIE DOG	Genus *Cynomys*
PRAIRIE DOG, BLACK-TAILED	*Cynomys ludovicianus*
PRONGHORN	*Antilocapra americana*
PUMA	*Felis concolor*
RABBIT	Family Leporidae
RABBIT, AMERICAN JACK	Genus *Lepus,* several species
RABBIT, DOMESTIC	*Oryctolagus cuniculus*
RABBIT, EUROPEAN	*Oryctolagus cuniculus*
RABBIT, WHITE-TAILED JACK	*Lepus townsendii*
RACCOON	Genus *Procyon*

RACCOON, NORTH AMERICAN	*Procyon loton*
RAT, FISHING	Genera *Ichthyomys, Rheomys,* and others
RAT, KANGAROO	Genus *Dipodomys*
RAT, LABORATORY	*Rattus norvegicus*
RAT, NORWAY	*Rattus norvegicus*
RAT, RICE	Genus *Oryzomys*
RAT, WOOD	Genus *Neotoma*
RATEL	*Mellivora capensis*
REINDEER	*Rangifer tarandus*
RHINOCEROS	Family Rhinocerotidae
RHINOCEROS, BLACK	*Diceros bicornis*
RORQUAL	Genus *Balaenoptera*
SEAL	Family Phocidae
SEAL, ALASKA FUR	*Callorhinus ursinus*
SEAL, CRAB-EATER	*Lobodon carcinophagus*
SEAL, ELEPHANT	Genus *Mirounga*
SEAL, GRAY	*Halichoerus grypus*
SEAL, HARP	*Pagophilus groenlandicus*
SEAL, HOODED	*Cystophora cristata*
SEAL, LEOPARD	*Hydrurga leptonyx*
SEAL, SOUTHERN ELEPHANT	*Mirounga leonina*
SEAL, WEDDELL'S	*Leptonychotes weddelli*
SEA LION, FALKLAND ISLAND	*Artocephalus australis*
SHEEP, BIGHORN	*Ovis canadensis*
SHEEP, DOMESTIC	*Ovis aries*
SHREW	Family Soricidae
SHREW, ETRUSCAN	*Suncus etruscus*
SHREW, LONG-TAILED	Genus *Sorex*
SHREW, OTTER	Family Potamogalidae
SHREW, PIGMY	*Microsorex hoyi*

APPENDIX

SHREW, SHORT-TAILED	*Blarina brevicauda*
SHREW, WATER	Genus *Neomys, Sorex palustris* and others
SKUNK	Subfamily Mephitinae
SKUNK, SPOTTED	*Spilogale putorius*
SKUNK, STRIPED	*Mephitis mephitis*
SLOTH, THREE-TOED	Genus *Bradypus*
SLOTH, TWO-TOED	Genus *Choloepus*
SPRINGBUCK	*Antidorcas marsupialis*
SQUIRREL	Family Sciuridae
SQUIRREL, ARCTIC GROUND	*Spermophilus undulatus*
SQUIRREL, CALIFORNIA GROUND	*Spermophilus beecheyi*
SQUIRREL, EUROPEAN RED	*Sciurus vulgaris*
SQUIRREL, FOX	*Sciurus niger*
SQUIRREL, GRAY	*Sciurus carolinensis*
SQUIRREL, MEXICAN GROUND	*Spermophilus mexicanus*
SQUIRREL, NORTH AMERICAN RED	*Tamiasciurus hudsonicus*
SQUIRREL, TASSEL-EARED	*Sciurus aberti*
TAMANDUA	*Tamandua tetradactyla*
TAPIR	Genus *Tapirus*
TELEDU	*Mydaus javanensis*
TENREC, MALAGASAY	*Tenrec ecaudatus*
TIGER	*Leo tigris*
TUCO-TUCO	Genus *Ctenomys*
VICUNA	*Vicugna vicugna*

WALLABY	Subfamily Macropodinae, several genera
WALRUS	*Odobenus rosmarus*
WAPITI	*Cervus canadensis*
WEASEL	Genus *Mustela*
WHALE	Order Cetacea
WHALE, BALEEN	Suborder Mysticeti
WHALE, BLUE	*Balaenoptera musculus*
WHALE, BOTTLENOSE	Genus *Hyperoodon*
WHALE, BOWHEAD	*Balaena mysticetus*
WHALE, FIN	*Balaenoptera borealis*
WHALE, GRAY	*Eschrichtius gibbosus*
WHALE, HUMPBACK	*Megaptera novaeangliae*
WHALE, RIGHT	Family Balaenidae
WHALE, SEI	*Balaenoptera acutorostrata*
WHALE, SPERM	*Physeter catodon*
WHALE, TOOTHED	Suborder Odontoceti
WILDEBEEST	Genus *Connochaetes*
WOLVERINE	*Gulo gulo*
WOLF	Genus *Canis*, several species
WOLF, TIMBER	*Canis lupus*
WOLF, TASMANIAN	*Thylacinus cynocephalus*
WOODCHUCK	*Marmota monax*
WOODRAT	Genus *Neotoma*
YAK	*Bos grunniens*
YAPOK	*Chironectes minimus*
ZEBRA	Genus *Equus*, several species

🐾🐾🐾 REFERENCE NOTES

CHAPTER 2 · BIRTH, GROWTH, DEVELOPMENT

1. J. Z. Young, *The Life of Vertebrates* (2nd ed.; New York: Oxford University Press, 1962), p. 553.
2. S. A. Asdell, "Reproduction and Development," in William V. Mayer and Richard G. Van Gelder, *Physiological Mammalogy* (New York: Academic Press, 1965) II, 31–32.
3. The figure of 18 feet high at the shoulder is generally given for *Baluchitherium,* based on reconstructions made at The American Museum of Natural History. I have been told that this height is probably erroneous, being based on several different species, none of which probably attained so great a size.
4. E. F. Adolph, "Ontogeny of Physiological Regulations in the Rat," *Quarterly Review of Biology,* XXXII (1957) 1, 89–137.

CHAPTER 3 · DISPERSAL

1. Carl B. Koford, "The Vicuña and the Puna," *Ecological Monographs,* XXVII (1957) 2, 153–219.
2. Jean M. Linsdale, *The California Ground Squirrel* (Los Angeles: University of California Press, 1946), p. 358.
3. Jean M. Linsdale and Lloyd P. Tevis, Jr., *The Dusky-Footed Wood Rat* (Los Angeles: University of California Press, 1951), pp. 420–436.

4. Rexford D. Lord, "Mortality Rates of Cottontail Rabbits," *Journal of Wildlife Management,* XXV (1961) 1, 33–40.

5. United Nations, *Demographic Yearbook 1961* (New York: United Nations Department of Economic and Social Affairs, 1961), pp. 216–262.

6. A. A. Lindsay, "Notes on the Crab-eater Seal," *Journal of Mammalogy,* XIX (1938) 4, 459–461.

7. A. Brazier Howell, *Aquatic Mammals* (Springfield: Charles C Thomas, 1930), pp. 24–47.

8. The speeds of mammals were obtained from various sources, including summaries in E. Lendell Cockrum, *Introduction to Mammalogy* (New York: Ronald Press Co., 1962), A. Brazier Howell, *Speed in Animals* (Chicago: University of Chicago Press, 1944); and T. G. Lang and K. S. Norris, "Swimming Speed of a Pacific Bottlenose Porpoise," *Science,* CLI (February 4, 1966), 588–590.

9. William J. Hamilton, Jr., American Mammals (New York: McGraw-Hill Book Co., 1939), pp. 254-255.

CHAPTER 4 · HOME RANGE, TERRITORY, SHELTER

1. W. Frank Blair, "Home Ranges and Populations of the Meadow Vole in Southern Michigan," *Journal of Wildlife Management,* IV (1940) 2, 149–161.

2. P. M. Youngman, "A Population of the Striped Field Mouse, *Apodemus agrarius coreae* in Central Korea," *Journal of Mammalogy,* XXXVII (1956) 1, 1–10.

3. W. Frank Blair, "Populations of the Deer-mouse and Associated Small Mammals in the Mesquite Association of Southern New Mexico," *Contributions of the Laboratory of Vertebrate Biology,* XXI (1943), 1–40.

4. Ernest P. Walker, *Mammals of the World* (Baltimore: Johns Hopkins Press, 1964), I, p. 175.

CHAPTER 5 · AIR, WATER, FOOD

1. Data obtained from various sources, including E. L. Cockrum, *Introduction to Mammalogy* (New York: Ronald Press,

1962), William S. Spector, *Handbook of Biological Data* (Philadelphia: W. B. Saunders Co., 1956), and L. Irving, "Respiration in Diving Mammals," *Physiological Reviews,* XIX (1939), 112–134.

2. E. F. Adolph, "Ontogeny of Physiological Regulations in the Rat," *Quarterly Review of Biology,* XXXII (1957), 89–137.

3. George B. Schaller, *The Mountain Gorilla* (Chicago: University of Chicago Press, 1963), p. 169.

4. Robert M. Chew, "Water Metabolism of Mammals," in William V. Mayer and Richard G. Van Gelder, *Physiological Mammalogy* (New York: Academic Press, 1965), II, 43–178.

5. E. J. Slijper, *Whales* (New York: Basic Books, 1962).

6. K. G. Brown, "Observations on the Newly Born Leopard Seal," *Nature,* CLXX (1952), 982–983.

CHAPTER 6 · DEFENSE AND PROTECTION

1. Lee R. Dice, "Effectiveness of Selection by Owls on Deer Mice (*Peromyscus maniculatus*) which Contrast in Color with Their Background," *Contributions of the Laboratory of Vertebrate Biology,* XXXIV (1947), 1–20.

2. Ralph S. Palmer, *The Mammal Guide* (Garden City: Doubleday and Co., 1954), p. 55.

3. John Eric Hill, "A Supposed Adaptation Against Sun Stroke in African Diurnal Rats," *Journal of Mammalogy,* XXIII (1942) 2, p. 210.

4. Charles Kayser, "Hibernation," in William V. Mayer and Richard G. Van Gelder, *Physiological Mammalogy* (New York: Academic Press, 1965), II, 179–296.

CHAPTER 7 · SOCIAL STRUCTURE AND POPULATION

1. Irwin Katz, "Behavioral Interactions in a Herd of Barbary Sheep (*Ammotragus lervia*)," *Zoologica,* XXXIV (1949) 1, 9–18.

2. C. R. Carpenter, "A Field Study of the Behavior and Social Relations of Howling Monkeys (*Alouatta palliata*)," *Comparative Psychology Monographs,* X (1934) 2, 1–168.

3. Raymond Pearl, "On Biological Principles Affecting Populations: Human and Other," *American Naturalist,* LXXI (1937), 50–68.

4. J. J. Christian, "Endocrine Adaptive Mechanisms and the Physiologic Regulation of Population Growth," in William V. Mayer and Richard G. Van Gelder, *Physiological Mammalogy* (New York: Academic Press, 1963), I, 189–353.

5. D. A. MacLulich, "Fluctuations in the Numbers of the Varying Hare," *University of Toronto Studies, Biological Series,* XLIII (1937), 1–136.

CHAPTER 8 · MATING, REPRODUCTION, GESTATION

1. S. A. Asdell, *Patterns of Mammalian Reproduction* (Ithaca: Comstock Publishing Co., Inc., 1946), p. 142.

2. *Ibid.,* p. 395.

3. *Ibid.,* p. 134.

CHAPTER 9 · THE VALUE OF MAMMALS

1. W. C. Allee, Alfred E. Emerson, Orlando Park, Thomas Park, and Karl P. Schmidt, *Principles of Animal Ecology* (Philadelphia: W. B. Saunders Co., 1949), p. 706.

2. Durward L. Allen and L. David Mech, "Wolves versus Moose on Isle Royale," *National Geographic* CXXIII (1963), 200–219.

3. C. A. Reed, "Animal Domestication in the Prehistoric Near East," *Science* CXXX (1959), 1629–1639.

🐾 🐾 🐾 GLOSSARY

[Italicized words in definitions also have entries in alphabetical order.]

ABOMASUM   The last of the four chambers of the ruminant "stomach"; the true stomach of a ruminant.

ADRENAL GLAND   A ductless gland consisting of an outer cortex and an inner medulla, that lies adjacent to the anterior end of each kidney. The medulla produces epinephrin and norepinephrin, which function in stress as well as in normal body function; the cortex produces several steroids, including cortisone, and functions in water metabolism, temperature regulation, and general metabolism.

ALIMENTAL MIGRATION   A periodic movement away from and back to a given area in search of food.

ALLANTOIS   An embryonic membrane that unites with the *chorion* to form the *placenta*.

ALTRICIAL   Having young that are relatively poorly developed at birth, usually with scant hair, the eyes closed, and little locomotor ability.

AMNION   The innermost membrane, filled with watery fluid, that encloses the developing embryo.

ANESTRUS   The quiescent period of the female reproductive cycle, followed by the *estrus*.

AXILLA   The armpit.

BACULUM   A bone in the *penis* of many kinds of mammals; the *os penis*.

BALEEN   A horny substance hanging in overlapping plates from the upper jaws of mysticete whales and functioning to strain food from the water; whalebone.

BICORNUATE UTERUS   A type of *uterus* in which the two horns are fused for their lower two-thirds to make a single chamber, and in which there is a single entrance to the *vagina*.

BIPARTITE UTERUS   A type of *uterus* in which the lower parts of each of the horns are fused together and in which there is a single entrance to the *vagina*.

BLASTOCYST   A stage of development of an egg after fertilization and cleavage, but before implantation in the wall of the *uterus*.

BRACHIATION   A form of locomotion that involves swinging from hand to hand along the underside of tree limbs.

BULBOURETHRAL GLAND   A secretory organ at the base of the *penis* that empties into the urethra a fluid that assists in the transport of sperm; *Cowper's gland*.

CAECUM   A blind pouch or saclike extension of the digestive tract.

CALORIE   The amount of heat required to raise one kilogram of water one degree Centigrade at 15°C. This is actually the large, or kilogram-calorie, equal to 1,000 small, or gram-calories, and equal to one-thousandth of a mega-calorie, or therm.

CARBOHYDRATE   A compound of carbon, hydrogen, and oxygen, the last two usually in a ratio of two to one. Sugars and starches. The primary sources of energy for animals.

CARNASSIAL TEETH   A pair of enlarged teeth modified for shearing in the Order Carnivora, usually the last upper premolar and the first lower molar.

CARUNCLE   A small, fleshy growth.

CERVIX   The junction of the *uterus* and the *vagina;* a neck.

CHORION   The outermost membrane surrounding the developing embryo. Together with the *allantois* it forms the *placenta*.

CLIMATIC MIGRATION   A periodic migration away from an area of one climate to an area of another climate and the return.

GLOSSARY

CLOACA   A chamber that receives the products of the digestive, excretory, and reproductive organs.

COPROPHAGY   The eating of dung.

CORPUS LUTEUM   A gland formed from the cells of the ovarian follicle after the release of the *ovum,* that produces secretions (*progestins*) that function in pregnancy and birth.

COWPER'S GLAND   The *bulbourethral gland.* Named for the British anatomist, William Cowper (1666–1709).

CURSORIAL   Adapted for running.

DECIDUOUS DENTITION   The first set of teeth, subsequently replaced by the permanent dentition.

DIGESTION   The process of preparing food for absorption and assimilation.

DIPHYODONT   Having two sets of teeth, the first being the milk teeth and the second the permanent teeth.

DISTANCE THRESHOLD   The least distance that an animal will permit another to approach without reacting.

DOMESTIC MAMMAL   A population of mammals that, through direct selection by man, has certain inherent morphological, physiological, or behavioral characteristics by which it differs from its ancestral stock.

DOMESTICATED MAMMAL   Any individual that has been docited, tamed, or made tractable.

DUPLEX UTERUS   A type of *uterus* in which each side or horn is completely separate from *oviduct* to entrance to the *vagina.*

EGG TOOTH   A horny projection on the tip of the snout used by hatchlings to help open the egg.

EMIGRATION   The act of leaving an area.

EPIDIDYMIS   The efferent tubules of the *testis;* the minute tubules in which the sperm are gathered from the testis.

ESTIVATION   A condition of torpor occurring in the summer.

ESTROGEN   Any of the hormones functioning in the production of *estrus.*

ESTRUS   A period in the reproductive cycle of the female when the animal is most physiologically and psychologically receptive to the male; the heat.

FAT   A compound of carbon and hydrogen with less oxygen than carbohydrates. Oils, greases, and waxes, either fluids or solids. The main chemicals of storage of energy in animals.

FOLLICLE   A minute cellular sac.

FOLLICULAR STIMULATING HORMONE (F.S.H.)   A secretion from the anterior *pituitary gland* that stimulates the development of *ova* in the ovary.

GALACTOGEN   A hormone from the *pituitary gland* that stimulates enlargement of the *mammary glands* during pregnancy and milk production after it.

GAMETIC MIGRATION   The periodic movement away from and back to an area for purposes of reproduction.

GESTATION   The period from conception to birth.

GLAND   An organ of secretion or excretion.

GLUCOSE   A simple sugar, generally the end-product of the breakdown of more complex compounds in digestion.

GLYCOGEN   A *carbohydrate* stored in muscles and liver; "animal starch."

GONAD   A reproductive organ in which *ova* or *sperm* are produced: an *ovary* or *testis*.

HEAT   The *estrus* in females or period of sexual excitement in the male, the rut.

HETEROTHERMIC   Having a variable body temperature that fluctuates with the environmental temperatures; cold-blooded.

HIBERNATION   A condition of deep torpor accompanied by great reduction of metabolic activity and little or no thermoregulation undergone by some mammals during the winter.

HOME RANGE   That area over which an animal roams in the course of its normal activities.

HOMOIOTHERMIC   Having the metabolic ability to maintain the body temperature relatively constant over a wide range of environmental temperatures; warm-blooded.

HORMONE   A chemical secreted by cells or ductless glands and carried in the blood stream; functions in regulation or coordination of chemical reactions in the body.

HYPSODONT   Referring to teeth having a high crown, as those of horses.

IMMIGRATION   The act of moving into a new area.

GLOSSARY

INGUINAL   Pertaining to the groin.

INSTINCT   An activity trait that is inherited.

INSTINCTIVE BEHAVIOR   Activity traits that are inherited and not learned.

KRILL   Planktonic crustaceans, mainly euphausiid shrimp.

LEARNED BEHAVIOR   Activity traits that are learned and not inherited.

LETHAL GENE   A unit of a chromosome that, when inherited from both parents, causes the death of the offspring.

LUTEINIZING HORMONE (L.H.)   A secretion from the anterior *pituitary gland* that stimulates the development of *ova* in the ovarian follicle and causes the follicle to rupture.

MAMMARY GLAND   The milk-producing organ of mammals, inactive in males.

MENOPAUSE   The cessation of periodic menstruation and ovulation; the end of reproductive ability in females.

METABOLISM   The sum of the activities, mainly chemical, of construction and destruction that occur within living organisms.

MIDDEN   A refuse heap.

MIGRATION   A periodic movement away from and back to a given area.

MILK DENTITION   The first set of teeth in mammals that have two sets.

MILK LINE   A pair of ridges along the trunk on each side of mammalian embryos on which the *mammary glands* develop.

MONESTRUS   Having only one heat period per year.

OMASUM   The third of the four large digestive chambers of the *ruminant* "stomach."

OS PENIS   A bone in the *penis* of many kinds of mammals; the *baculum*.

OVARY   The female *gonad* in which the eggs are formed and nourished.

OVIDUCT   The tube through which eggs pass from the *ovary* to the *uterus*.

OVUM   An egg; the female sex cell, or gamete.

PECK ORDER   A hierarchal social structure in which the more dominant animals peck the less dominant; now used for almost any hierarchal social structure.

PECTORAL   Pertaining to the chest or thoracic area.

PENIS   The male copulatory organ.

PHOTOPERIOD   The duration of daylight in relation to the duration of darkness over a 24-hour period.

PHYSIOGRAPHY   Physical geography.

PITUITARY   A gland at the base of the brain; the hypophysis. The anterior portion or lobe produces hormones that affect growth, reproduction, milk production, the adrenal glands, the thyroid glands, and blood sugar. The posterior lobe produces hormones that affect smooth muscle, and blood pressure.

PLACENTA   An organ by which the embryo is attached to the *uterus* of the mother and through which metabolic products pass.

PLANKTON   Marine organisms, mainly small, that float and are moved passively by action of wind, waves, or current.

POLYESTRUS   Having more than one heat period per year.

POPULATION   A group of individuals within a specified area of time and space; the organisms, collectively, inhabiting an area or region.

PRECOCIAL   Having young that are relatively well developed at birth, usually with much hair, the eyes open, and good locomotor ability.

PREHENSILE   Adapted for grasping or holding.

PROGESTIN   An ovarian hormone, produced by the *corpus luteum,* that helps to prepare the *uterus* for receiving the fertilized egg.

PROLACTIN   A hormone from the anterior *pituitary* that functions in causing development of the *mammary glands* and the production of milk.

PROSTATE   A gland at the lower end of the *vas deferens* and around the bladder that secretes some of the fluids that make up *semen.*

PROTEIN   A compound of carbon, hydrogen, oxygen, and nitrogen, and usually small amounts of sulfur and other elements.

GLOSSARY

Amino acids. The primary source of materials for growth in animals.

PSEUDOHERMAPHRODITISM   A condition in which an animal that is genetically of one sex has some organs and shows some resemblances to the other.

PSEUDOPREGNANCY   A false pregnancy.

RELAXIN   A hormone produced by the *ovary* that causes the ligaments of the pelvis to relax at parturition.

RETICULUM   The second of the four large digestive chambers of the ruminant "stomach."

RUMEN   An esophageal pouch of certain artiodactyls; the first of the four large digestive chambers of the *ruminant* "stomach."

RUMINANT   A herbivorous mammal that regurgitates and chews a cud and that has special modifications of the digestive tract for this process.

RUT   The period of sexual excitement in males.

SCROTUM   A skin-covered sac, actually an extension of the abdominal cavity, that is suspended below the anus and which contains the *testes*.

SECONDARY SEXUAL CHARACTER   External differences, other than the reproductive organs, between the sexes.

SEMEN   The fluid containing *spermatozoa* and glandular secretions.

SEMINAL VESICLE   A gland at the lower end of the *vas deferens* that secretes some of the fluids that make up semen.

SESSILE   Not ambulatory; fixed to one spot.

SIGMOID   Shaped like the letter S.

SIMPLEX UTERUS   A type of uterus in which both horns are completely fused together to form a single chamber.

SPERMATOZOA   The matured male sex cells or gamete.

SUPERFETATION   The condition of being pregnant with two litters at the same time.

TERRITORY   An area, generally surrounding the home, that is defended against other individuals of the same species.

TESTIS   The male gonad, in which *spermatozoa* are formed.

TESTOSTERONE   A hormone, secreted by the *testis,* that functions in causing the development of the male secondary sexual characteristics and accessory sex organs.

TORPIDITY The condition of being devoid of the power of movement or feeling; sluggish in function or action.

URETHRA The duct through which urine is discharged from the bladder to the outside; in males it also carries *semen*.

UROGENITAL SINUS A cavity into which the products of the excretory and reproductive systems empty.

UTERUS The enlarged posterior portion of an oviduct in which embryos are retained and nourished; the womb.

VAGINA The terminal portion of the female reproductive tract; the receptacle for the copulatory organ of the male in mating.

VAGINAL SINUS A cavity in marsupials, formed from the two *vaginas,* in which the development of the fetus takes place.

VAS DEFERENS A tube for sperm transport, extending from the *epididymis* to the *urethra*.

VIVIPAROUS Producing young from eggs retained in the mother's body and nourished by her blood supply.

WINTER DORMANT An animal that is inactive, but not torpid, in winter.

ZYGOTE A fertilized egg.

🐾🐾🐾 SELECTED BIBLIOGRAPHY

Allee, W. C., A. E. Emerson, O. Park, T. Park, and K. P. Schmidt. *Principles of Animal Ecology.* Philadelphia: W. B. Saunders Co., 1949.
    An excellent compendium of principles and examples.

Allen, G. M. *Bats.* Cambridge, Mass.: Harvard University Press, 1939.
    A very good survey, although lacking modern information on echolocation.

————. "A Checklist of African Mammals," *Bulletin of the Museum of Comparative Zoology,* Harvard University, LXXXIII (1939), 1–763.
    The only complete checklist for Africa.

————. *Extinct and Vanishing Mammals of the Western Hemisphere with Marine Species of all Oceans.* New York: American Committee for International Wildlife Protection. Special Publication 11, 1942.
    Excellent historical coverage.

Anderson, S., and J. K. Jones, Jr., eds. *Recent Mammals of the World. A Synopsis of Families.* New York: The Ronald Press Company, 1967.
    A concise technical survey, especially good for zoogeography and characteristics of families.

Asdell, S. A. *Patterns of Mammalian Reproduction.* Ithaca: Cornell University Press, 1964.
    An extensive survey.

Bourliere, F. *The Natural History of Mammals.* New York: Alfred A. Knopf, Inc., 1964.
: A highly readable, well-illustrated account of the biology of mammals.

Brody, S. *Bioenergetics and Growth.* New York: Reinhold Publishing Corp., 1945.
: A technical treatise, concerned mainly with domestic animals.

Burrell, H. *The Platypus.* Sydney, Australia: Angus & Robertson, Ltd., 1927.
: The prime work on this species.

Burt, W. H., and R. P. Grossenheider. *A Field Guide to the Mammals.* Boston: Houghton Mifflin Company, 1964.
: Excellent paintings and maps; concise and limited biology.

Cabrera, A. "Catalogo de los Mamiferos de America del Sur," *Revista Museo Argentino de Ciencias Naturales, Ciencias Zoologicas,* IV (1958), Vol. 4, No. 1, 1–307; IV (1961), Vol. 4, No. 2, 308–732.
: The only complete checklist for South America (in Spanish).

Cabrera, A., and Yepes, J. *Mamiferos Sud-Americanos.* Buenos Aires, Argentina: Co. Argentina Edit., 1940.
: The only complete account of the mammals of South America (in Spanish).

Cahalane, V. H. *Mammals of North America.* New York: The Macmillan Company, 1947.
: Accurate and readable life histories.

Carpenter, C. R. "A Field Study of the Behavior and Social Relations of the Howling Monkeys," *Comparative Psychology Monographs,* X, No. 2 (1934), 1–168.
: A classic first field study.

Carrington, R. *The Mammals.* New York: Time-Life Books, 1963.
: A well-illustrated, accurately written life-history survey.

Chasen, F. N. "A Handlist of Malaysian Mammals," *Bulletin of the Raffles Museum,* XV (1940), 1–209.
: A checklist.

Cockrum, E. L. *Introduction to Mammalogy.* New York: The Ronald Press Company, 1962.
: A textbook for colleges.

Colbert, E. H. *Evolution of the Vertebrates.* New York: John Wiley & Sons, Inc., 1955.
: An authoritative work, about half of it concerned with mammals.

SELECTED BIBLIOGRAPHY

Couturier, M. A. J. *L'Ours Brun.* Grenoble, France: M. A. J. Couturier, 1954.
> Extensive coverage of the brown bear (in French).

Crandall, L. S. *The Management of Wild Mammals in Captivity.* Chicago: University of Chicago Press, 1964.
> Thorough and detailed coverage of mammals of the world and their habits in relation to captivity.

Darling, F. F. *A Herd of Red Deer. A Study in Animal Behaviour.* Oxford, England: Oxford University Press, 1937.
> A classic study.

Davis, D. E., and F. B. Golley. *Principles in Mammalogy.* New York: Reinhold Publishing Corp., 1963.
> A textbook exemplifying biological principles with mammals.

Darlington, P. J., Jr. *Zoogeography: The Geographical Distribution of Animals.* New York: John Wiley & Sons, Inc., 1957.
> An important source book.

Ellerman, J. R., and T. C. S. Morrison-Scott. *Checklist of Palaearctic and Indian Mammals, 1758 to 1946.* London: British Museum of Natural History, 1951.
> The major checklist for the area.

Ellerman, J. R., T. C. S. Morrison-Scott, and R. W. Hayman. *Southern African Mammals, 1758–1951: A Reclassification.* London: British Museum of Natural History, 1953.
> A checklist, with some keys, for southern Africa.

Elton, C. *Voles, Mice, and Lemmings.* Oxford: Clarendon Press, 1942.
> Emphasizes populations studies.

Flower, W. H., and R. Lydekker. *An Introduction to the Study of Mammals, Living and Extinct.* London: Adam and Charles Black, 1891.
> Mainly anatomical in approach; somewhat outdated in biology and taxonomy.

Grasse, P., ed. *Traite de Zoologie.* Vol. 17, *Mammiferes. Les Ordres: anatomie, ethologie, systematique.* Paris: Masson et Cie., 1955.
> An encyclopedic work (in French).

Gray, J. *How Animals Move.* Cambridge, England: Cambridge University Press, 1953.
> An excellent elementary introduction to locomotion.

Griffin, D. R. *Listening in the Dark.* New Haven, Conn.: Yale University Press, 1958.
> On bat echolocation.

Hafez, E. S. E., ed. *The Behaviour of Domestic Animals.* London: Baillere, Tindall, and Cox, 1962.
  Contains much interesting basic behavioral information.
Hall, E. R., and K. R. Kelson. *The Mammals of North America.* New York: The Ronald Press Company, 2 vols., 1959.
  Keys and distribution maps for all species of mammals north of Panama, although little information on biology.
Hamilton, W. J., Jr. *American Mammals, Their Lives, Habits, and Economic Relations.* New York: McGraw-Hill Book Company, 1939.
  An excellent work, but unfortunately outdated.
Harper, F. *Extinct and Vanishing Mammals of the Old World.* New York: American Committee for International Wildlife Protection. Special Publication 12, 1945.
  Companion volume to Allen, C. M., *Extinct and Vanishing Mammals of the Western Hemisphere,* 1942.
Hartman, C. G. *Possums.* Austin: University of Texas Press, 1952.
  A popular account of the discovery and life history of the North American opossum.
Henderson, J., and E. L. Craig. *Economic Mammalogy.* Springfield, Ill.: Charles C Thomas, 1932.
  A very fine work, although now hopelessly out of date.
Howell, A. B. *Aquatic Mammals.* Springfield, Ill.: Charles C Thomas, 1930.
  Unusually good discussion of anatomical adaptations for aquatic life.
Iredale, T., and E. Troughton. "A Check-list of the Mammals Recorded from Australia," *Australian Museum Memoir,* VI (1934), 1–122.
  The most recent complete list for the area, now quite outdated.
King, J. E. *Seals of the World.* London: British Museum of Natural History, 1964.
  A superb blend of the technical and the popular.
Laurie, E. M. O., and J. E. Hill. *List of Land Mammals of New Guinea, Celebes and Adjacent Islands,* 1758–1952. London: British Museum of Natural History, 1954.
  A checklist.
Mayer, W. B., and R. G. Van Gelder, eds. *Physiological Mammalogy.* New York: Academic Press Inc., 2 vols., 1963, 1965.
  Contains articles on populations, adreno-pituitary stress, reproduction, water metabolism, and hibernation.

SELECTED BIBLIOGRAPHY

Miller, G. S., Jr., and R. Kellogg. "List of North American Recent Mammals," *Bulletin of the U.S. National Museum,* CCV (1955), 1–954.
    A checklist.
Morris, D. *The Mammals, A Guide to the Living Species.* New York: Harper & Row, 1965.
    Lists all the genera and most of the species of mammals and includes general information and photographs for about 200 of them.
Murie, A. "The Wolves of Mount McKinley," *Fauna of the National Parks of the United States, Fauna Series,* V (1944), 1–238.
    A fine account of the life history of wolves.
Murie, O. J. *The Elk of North America.* Harrisburg, Pa.: The Stackpole Co., 1951.
    Original observations combined with a compilation of the literature.
Norman, J. R., and F. C. Fraser. *Giant Fishes, Whales, and Dolphins.* London: Putnam & Co., Ltd., 1948.
    Good descriptive material on cetaceans.
Odum, E. P. *Fundamentals of Ecology.* Philadelphia: W. B. Saunders Co., 1953.
    A very good textbook.
Palmer, R. S. *The Mammal Guide, Mammals of North America North of Mexico.* Garden City, N.Y.: Doubleday & Company, Inc. 1954.
    Concise descriptions and life histories—excellent.
Prater, S. H. *The Book of Indian Animals.* Bombay, India: Bombay Natural History Society, 1948.
    Description, distribution, and habits concisely outlined.
Prosser, C. L., ed. *Comparative Animal Physiology.* Philadelphia: W. B. Saunders Co., 1950.
    A detailed technical text.
Roberts, A. *The Mammals of South Africa.* Johannesburg, South Africa: Central News Agency, 1951.
    A well-illustrated, detailed account, unsatisfactory as to classification.
Romer, A. S. *Vertebrate Paleontology.* Chicago: University of Chicago Press, 1945.
    An excellent textbook, now revised.

Schaller, G. B. *The Mountain Gorilla, Ecology and Behavior*. Chicago: University of Chicago Press, 1963.
    A fine field study.
Scheffer, V. B. *Seals, Sea Lions, and Walruses*. Stanford, Calif.: Stanford University Press, 1958.
    A good source, but difficult to use.
Scott, J. P. *Animal Behavior*. Chicago: University of Chicago Press, 1958.
    A very good introduction to the subject.
Simpson, G. G. "The Principles of Classification and a Classification of Mammals," *Bulletin of the American Museum of Natural History,* 85 (1945), 1–350.
    This very useful article has stood the test of time well.
Slijper, E. J. *Whales*. New York: Basic Books, Inc., 1962.
    A readable summary of knowledge about whales.
Tate, G. H. H. *Mammals of Eastern Asia*. New York: The Macmillan Company, 1947.
    A useful handbook.
Taylor, W. P., ed. *The Deer of North America*. Harrisburg, Pa.: The Stackpole Co., 1956.
    A useful work, oriented toward game management.
Troughton, E. *Furred Animals of Australia*. New York: Charles Scribner's Sons, 1947.
    Mainly descriptive and taxonomic.
Van den Brink, F. H. *Die Saugetiere Europas*. Hamburg, West Germany: Parey, 1956.
    A field guide with maps and color plates (in German).
Walker, E. P., *et. al. Mammals of the World*. Baltimore: The Johns Hopkins Press, 3 vols., 1964.
    The first two volumes illustrate and outline life histories of all genera of mammals; the third is an almost unusable bibliography.
Whitney, L. F., and A. B. Underwood. *The Raccoon*. Orange, N. J.: Practical Science Publishing Co., 1952.
    Summarizes the life-history information.
Wynne-Edwards, V. C. *Animal Dispersion in Relation to Social Behavior*. London: Oliver and Boyd, 1962.
    A useful sourcebook.
Zeuner, F. E. *A History of Domesticated Animals*. New York: Harper & Row, 1963.
    Archaeological rather than biological orientation.

SELECTED BIBLIOGRAPHY

JOURNALS

Journal of Mammalogy
　Articles in English. Excellent. Each issue lists "recent literature." Quarterly.

Journal of Wildlife Management
　Articles in English. Orientation is toward management and techniques. Quarterly.

Mammalia
　Most articles in French, but some in English. Quarterly.

Zeitschrift für Saugetierkunde
　Most articles in German, but some in English and French. Many articles on behavior. Quarterly.

Zoological Record
　A cross-indexed listing of the literature of zoology. Annual.